OUTLINES OF
THE THEORY OF
ELECTROMAGNETISM

OUTLINES

OF

THE THEORY OF ELECTROMAGNETISM

A SERIES OF LECTURES DELIVERED BEFORE
THE CALCUTTA UNIVERSITY

by

GILBERT T. WALKER, M.A., Sc.D., F.R.S.

DIRECTOR-GENERAL OF OBSERVATORIES, INDIA,
AND FORMERLY FELLOW OF TRINITY COLLEGE, CAMBRIDGE

Cambridge :
at the University Press
1910

CAMBRIDGE
UNIVERSITY PRESS

University Printing House, Cambridge CB2 8BS, United Kingdom

Cambridge University Press is part of the University of Cambridge.

It furthers the University's mission by disseminating knowledge in the pursuit of education, learning and research at the highest international levels of excellence.

www.cambridge.org
Information on this title: www.cambridge.org/9781316619803

First published 1910
First paperback edition 2016

A catalogue record for this publication is available from the British Library

ISBN 978-1-316-61980-3 Paperback

PREFACE

THE University of Calcutta did me the honour early in 1908 to appoint me Reader, and asked me to deliver a series of lectures upon some subject, preferably electrical, which would be of use to the lecturers in the outlying colleges as well as to the more advanced students in Calcutta. It was a condition of the appointment that the lectures should subsequently be published, and it appeared that I could best attain these ends by attempting to put some of the more important developments of electromagnetic theory into a connected and convenient form. It is therefore chiefly in the mode of presentation, rather than in the subject matter, that any originality which the lectures may possess must be sought.

For the material I am very largely indebted to the writings of H. A. Lorentz, while some features in the treatment of vector analysis are taken from the *Vector Analysis* of E. B. Wilson.

G. T. W.

October, 1910.

CONTENTS

CHAPTER I

VECTOR ANALYSIS

CHAPTER II

APPLICATIONS OF VECTORIAL METHODS TO MAGNETOSTATICS

CHAPTER III

THE THEORY OF MAXWELL AS EXPRESSED BY HERTZ

CHAPTER IV

HERTZ'S EQUATIONS FOR MOVING MEDIA

CHAPTER V

SOME EFFECTS DUE TO THE MOTION OF CHARGED PARTICLES THROUGH A STATIONARY AETHER

CHAPTER VI

THE ELECTRON THEORY OF LORENTZ APPLIED TO STATIONARY MEDIA

CHAPTER VII

THE ELECTRON THEORY OF LORENTZ APPLIED TO MOVING MEDIA

CHAPTER I.

VECTOR ANALYSIS.

1. WE may divide the quantities that we meet with in physics into two classes according as they have or have not a direction associated with them. Quantities of the former type which obey the parallelogram law, such as velocities and forces, are called *vectors*, while those of the latter type, such as time intervals, masses and temperatures, are called *scalars*. The algebra of scalars is that of ordinary real quantities and need not concern us further.

2. If the straight lines OP, OQ represent two vectors

The addition and subtraction of vectors.

A, B, we shall define $A + B$, the geometric sum, as represented by the diagonal OR of the parallelogram $POQR$. This is the same as $B + A$.

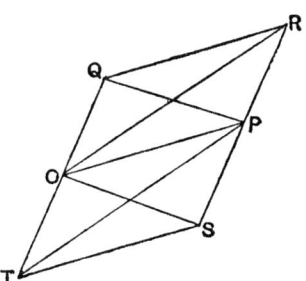

Similarly $A - B$ is the sum of the vectors OP, OT, where OT is equal and opposite to OQ. Thus $A - B$, the geometric difference, is represented by the diagonal OS of the parallelogram $TOPS$, i.e. by the second diagonal QP of the original parallelogram.

3. We now define i, j, k as vectors of unit length along rectangular axes OX, OY, OZ; so that if P be the point (x, y, z)

and PM, PN be perpendiculars to OX and the plane XOY respectively, the vectors OM, MN, NP will represent $\mathbf{i}x$, $\mathbf{j}y$, $\mathbf{k}z$.

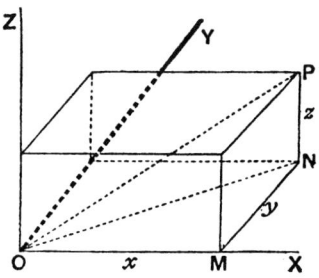

Now the vector ON, being the sum of the vectors OM, MN, will represent $\mathbf{i}x + \mathbf{j}y$; hence $\mathbf{i}x + \mathbf{j}y + \mathbf{k}z$ will be represented by the sum of the vectors ON, NP, or OP, which we shall denote by $\mathbf{r}$. Thus

$$\mathbf{r} = \mathbf{i}x + \mathbf{j}y + \mathbf{k}z.$$

If l, m, n be the direction cosines of OP we have $x = lr$, $y = mr$, $z = nr$; so

$$\mathbf{r} = r\,(\mathbf{i}l + \mathbf{j}m + \mathbf{k}n).$$

4. Consider a second line OP' defined by r', l', m', n'; and let the angle POP' be denoted by θ. Since the projection of the vector OP along OP' is equal to the sum of the projections of the component vectors OM, MN, NP along that line,

Illustration I.

$$OP \cos\theta = xl' + ym' + zn',$$

i.e.

$$r \cos\theta = rll' + rmm' + rnn'.$$

Hence

$$rr' \cos\theta = xx' + yy' + zz'$$

and

$$\cos\theta = ll' + mm' + nn'.$$

5. If $\mathbf{r}$, $\mathbf{r}'$ be two consecutive vectors OP, OP' at times t, $t + \delta t$ to a particle P moving with velocity $\mathbf{v}$,

Illustration II.

$$\mathbf{v} = \text{limit of } \frac{PP'}{\delta t}$$

$$= \text{limit of } \frac{\mathbf{r}' - \mathbf{r}}{\delta t}$$

$$= \text{limit of } \frac{\delta \mathbf{r}}{\delta t}$$

$$= \dot{\mathbf{r}}.$$

Similarly if $\mathbf{v}$, $\mathbf{v}'$ be the velocities at the times t, $t + \delta t$, we have the acceleration

$$\mathbf{f} = \text{limit of } \frac{\mathbf{v}' - \mathbf{v}}{\delta t}$$
$$= \dot{\mathbf{v}}$$
$$= \ddot{\mathbf{r}}.$$

6. Taking (r, θ) as the polar coordinates of a point P, let **Illustration** $\mathbf{R}$, $\mathbf{T}$ be unit vectors along and at right angles to **III.** OP in the directions r, θ increasing; then, if Q be the point $(1, \theta)$ which lies on OP and remains at unit distance from O as θ varies,

$\dot{\mathbf{R}} = $ the velocity of the point Q

   $= \dot{\theta}\mathbf{T}$, for its direction is that of $\mathbf{T}$, i.e. that of OS, at right angles to OP.

Similarly if OS remains of unit length,

   $\dot{\mathbf{T}} = $ the velocity of the point S

      $= -\dot{\theta}\mathbf{R}$, for its direction is that of QO.

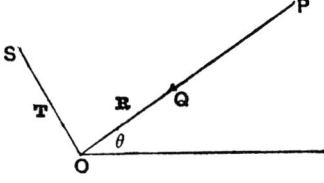

Now $\mathbf{r} = r\mathbf{R}$,

$$\therefore \ \mathbf{v} = \dot{\mathbf{r}} = \dot{r}\mathbf{R} + r\dot{\mathbf{R}}$$
$$= \dot{r}\mathbf{R} + r\dot{\theta}\mathbf{T}.$$

Thus the velocity is made up of $\dot{r}$ along OP and $r\dot{\theta}$ at right angles to it.

Further

$$\mathbf{f} = \dot{\mathbf{v}}_{s} = \frac{d}{dt}(\dot{r}\mathbf{R} + r\dot{\theta}\mathbf{T})$$
$$= \ddot{r}\mathbf{R} + \dot{r}\dot{\mathbf{R}} + \mathbf{T}\frac{d}{dt}(r\dot{\theta}) + r\dot{\theta}.\dot{\mathbf{T}}$$
$$= (\ddot{r} - r\dot{\theta}^2).\mathbf{R} + \frac{1}{r}\frac{d}{dt}(r^2\dot{\theta}).\mathbf{T},$$

giving the usual components along R and T. The acceleration of a particle whose three-dimensional polar coordinates are (r, θ, ϕ) may be obtained in a similar manner.

7. Let us consider the functions of the second degree with

Scalar and vector products of vectors.

which we are concerned in physics. If a force F act upon a particle moving with velocity u, the rate at which work is done is the product of the numerical scalar values of F and u, multiplied by the cosine of the angle between the directions of F and u. This is a scalar quantity, and so, if we have two vectors

$$\mathbf{r} \equiv \mathbf{i}x + \mathbf{j}y + \mathbf{k}z,$$
$$\mathbf{r}' \equiv \mathbf{i}x' + \mathbf{j}y' + \mathbf{k}z',$$

we give the title of their scalar product to $rr' \cos \theta$ or $xx' + yy' + zz'$. We shall denote it by $\{\mathbf{rr'}\}$ or $\{\mathbf{r'r}\}$; and when no ambiguity can arise from the omission of the brackets they will usually be omitted.

8. On the other hand if a force $(x'y'z')$ be applied at the point (x, y, z) the couple about the origin has components

$$yz' - y'z, \quad zx' - z'x, \quad xy' - x'y.$$

This is a vector $\mathbf{r}''$ of scalar magnitude $rr' \sin \theta$, and its direction is at right angles to r, r', being that of the axis about which r must be rotated in the right-handed direction in order to bring it into coincidence with r'. If r be due east, and r' due north, r'' is towards the zenith. The directions r, r', r'' form a right-handed system, and it is important to remember that the axes of reference OX, OY, OZ must always be chosen so as to form a right-handed system. The vector r'' is called the vector product of r, r', and is denoted by [rr'], these square brackets never being omitted. Thus

$$[\mathbf{rr'}] = \mathbf{i}\,(yz' - y'z) + \mathbf{j}\,(zx' - z'x) + \mathbf{k}\,(xy' - x'y)$$
$$= \begin{vmatrix} \mathbf{i}, & \mathbf{j}, & \mathbf{k} \\ x, & y, & z \\ x', & y', & z' \end{vmatrix}$$
$$= -[\mathbf{r'r}].$$

9. Its numerical magnitude is the area of the parallelogram whose sides are **r**, **r'**. Thus the scalar product

Lemma I.

C [AB] = (scalar magnitude of **C**) × (area of parallelogram **A, B**) × (cosine of the angle between **C** and the positive normal to **A, B**)

= volume of the parallelepiped of which three adjacent edges are **A, B, C**, being positive when **A, B, C** form a right-handed system.

If the components of **A** be A_1, A_2, A_3, &c., the components of [AB] being $A_2B_3 - A_3B_2$, &c., we have

$$C\,[AB] = \begin{vmatrix} C_1, & C_2, & C_3 \\ A_1, & A_2, & A_3 \\ B_1, & B_2, & B_3 \end{vmatrix}$$

$$= A\,[BC] = B\,[CA] \quad \ldots\ldots\ldots\ldots(1).$$

Lemma II. **10.** We have

$$[A\,[BC]] = \begin{vmatrix} i, & j, & k \\ A_1, & A_2, & A_3 \\ B_2C_3 - B_3C_2, & B_3C_1 - B_1C_3, & B_1C_2 - B_2C_1 \end{vmatrix}$$

$$= i\,(B_1 . \{CA\} - C_1 . \{BA\}) + j\,(B_2 . \{CA\} - C_2 . \{BA\})$$
$$+ k\,(B_3 . \{CA\} - C_3 . \{BA\})$$

$$= (iB_1 + jB_2 + kB_3)\,CA - (iC_1 + jC_2 + kC_3)\,BA$$

$$= B . CA - C . BA \quad \ldots\ldots\ldots\ldots\ldots(2).$$

11. Let us denote by ∇ the operator

Vectorial differentiation.

$$i\,\frac{d}{dx} + j\,\frac{d}{dy} + k\,\frac{d}{dz},$$

so that, if ϕ be any scalar function of x, y, z,

$$\nabla \phi = i\,\frac{d\phi}{dx} + j\,\frac{d\phi}{dy} + k\,\frac{d\phi}{dz}$$

$$= R, \text{ say.}$$

If the magnitude and direction of **R** be R and (l, m, n), we shall have

$$Rl = \frac{d\phi}{dx}, \quad Rm = \frac{d\phi}{dy}, \quad Rn = \frac{d\phi}{dz}.$$

Now if we consider the rate of change of ϕ in the direction of any unit vector **D** or (λ, μ, ν) we shall have, on going a small distance δs,

$$\delta\phi = \frac{d\phi}{dx} . \lambda \delta s + \frac{d\phi}{dy} . \mu \delta s + \frac{d\phi}{dz} . \nu \delta s$$

$$= Rl\lambda \delta s + Rm\mu \delta s + Rn\nu \delta s$$

$$= R \cos \theta \delta s,$$

where θ is the angle between **R** and **D**.

$\therefore \dfrac{d\phi}{ds} = R \cos \theta = $ **RD** and is a maximum when $\theta = 0$, i.e. in the direction of **R**. It is zero in a direction perpendicular to **R**. Thus **R** is along the normal to the surface $\phi = $ constant, and its scalar magnitude is $\dfrac{d\phi}{dn}$, where dn is an element of this normal.

We have seen that the rate of change of ϕ in any direction **D** is $\{$**D** $\nabla\phi\}$, the component of $\dfrac{d\phi}{dn}$ along **D**: hence $\nabla\phi$ is a vector which is independent of the selection of the axes.

12. Green's theorem tells us that for a region in which any vector **u** or (u, v, w) is finite, continuous and single-valued

Green's theorem.

$$\int dv \left(\frac{du}{dx} + \frac{dv}{dy} + \frac{dw}{dz}\right) = -\int dS \, (lu + mv + nw),$$

where (l, m, n) is the normal **N** of unit length drawn into the region.

Thus $\int dv \, \{\nabla \mathbf{u}\} = -\int dS \, \{\mathbf{N}\mathbf{u}\}$

$$= -\int \{\mathbf{dS}\, \mathbf{u}\},$$

if **dS** be treated as a vector whose direction is **N**.

13. If **u** be the velocity of a fluid, $-\int \{\mathbf{dS}\, \mathbf{u}\}$ is the rate at which fluid leaves the region: thus, applying the theorem to an element of volume, $\nabla\mathbf{u}$ is the rate at which the fluid expands per unit volume; hence its name of 'divergence' of **u**. It is usually written div **u**.

The operator 'divergence.'

14. If $u = B\phi$, where B is a vector and ϕ is a scalar, Green's theorem becomes

$$\int dv \{\nabla B\phi\} = -\int \{dS\ B\phi\}.$$

Now on the left side we may replace $\nabla B\phi$ by

$$\nabla_1 B\phi + \nabla_2 B\phi,$$

where in the first term ∇ operates on B only, and in the second on ϕ only: thus it becomes $\phi \{\nabla B\} + \{B\nabla\} \phi$. Hence the theorem

$$\int dv (\phi\operatorname{div} B + B\nabla . \phi) = -\int \{dS\ B\} \phi \quad \dots\dots\dots(3).$$

15. (a) If ϕ be a scalar quantity, integration, as in Green's theorem, gives

Analogues of Green's theorem.

$$\int dv \left(i\frac{d\phi}{dx} + j\frac{d\phi}{dy} + k\frac{d\phi}{dz} \right)$$
$$= -\int dS\,(il\phi + jm\phi + kn\phi),$$
$$\therefore \int dv\,\nabla\phi = -\int dS\,N\phi$$
$$= -\int dS\,\phi \quad\dots\dots\dots\dots\dots(4).$$

(b) If u be a vector whose components are (u, v, w),

$$\int dv\,[\nabla u] = \int dv \begin{vmatrix} i & , & j & , & k \\ \dfrac{d}{dx} , & \dfrac{d}{dy} , & \dfrac{d}{dz} \\ u & , & v & , & w \end{vmatrix}.$$

Now in integrating we replace $\int dv \dfrac{d}{dx}$ by $-\int dSl$, &c.; hence we get

$$-\int dS \begin{vmatrix} i, & j, & k \\ l, & m, & n \\ u, & v, & w \end{vmatrix},$$

or $\qquad\qquad -\int dS\,[Nu],$

or $\qquad\qquad -\int [dS\ u].$

We call $[\nabla u]$ the 'rotation' of u and write it rot u. Thus

The operator 'rotation.'

$$\int dv \operatorname{rot} u = -\int [dS\ u] \quad \dots\dots\dots\dots(5).$$

16. Putting $u = B\phi$, where B is a vector and ϕ a scalar, and replacing $[\nabla B\phi]$ by $[\nabla_1 B\phi] + [\nabla_2 B\phi]$, we obtain

$$\int dv \, (\phi \operatorname{rot} B - [B\nabla]\,\phi) = -\int [dS\,B]\,\phi \quad\ldots\ldots\ldots(6).$$

17. Stokes' theorem tells us that the line integral

Stokes' theorem. $\int (u\,dx + v\,dy + w\,dz)$ or $\int \{ds\,u\}$ round the margin of any area is equal to the surface integral over it,

$$\int dS \left(l\left(\frac{dw}{dy} - \frac{dv}{dz}\right) + m\left(\frac{du}{dz} - \frac{dw}{dx}\right) + n\left(\frac{dv}{dx} - \frac{du}{dy}\right) \right),$$

or $\qquad\qquad \int dS \, \{N \operatorname{rot} u\}, \quad\text{or}\quad \int \{dS \operatorname{rot} u\}.$

We may show that $\operatorname{rot} u$ has a meaning independent of the position of the axes exactly as we did in the case of $\nabla\phi$: for the line integral round an element of area dS is equal to the component normal to dS of $\operatorname{rot} u$: and the line integral is independent of the particular axes selected.

18. It may be of interest to have a proof of Stokes' theorem in terms of vector analysis.

Proof of Stokes' theorem by vector analysis. Let us consider one only, dS, of the elements into which the surface S may be divided; and let r be the vector joining a fixed point P_0 in this element to any point P which lies on its margin. Then if P' be a consecutive point $r + dr$, the area of the triangle $P_0 PP'$ will be equal in magnitude and direction to $\frac{1}{2}[r, dr]$. Thus

$$\{dS \,.\, \operatorname{rot} u\} = \tfrac{1}{2}\int [r, dr] \operatorname{rot} u,$$

the integration being round the margin of dS,

$$= \tfrac{1}{2}\int dr \, [\operatorname{rot} u, r]$$

$$= \tfrac{1}{2}\int dr \, [[\nabla_1 u]\, r],$$

where ∇_1 operates only on u,

$$= \tfrac{1}{2}\int dr \, (\{r\nabla\}\, u - \nabla_1 \{ur\}).$$

Now by Taylor's theorem, if squares of small quantities be neglected, the value of u at P will exceed its value u_0 at P_0 by $r\nabla \,.\, u.$ Thus

$$\int dr \, \{r\nabla\}\, u = \int dr \, (u - u_0) = \int dr \,.\, u - u_0 \int dr.$$

Also $\nabla_1\{ur\} = \nabla\{ur\} - \nabla_2\{ur\}$, where ∇_2 operates only on r; and $\nabla_2\{ur\} = \nabla_2(ux + vy + wz) = u$: thus we find on substitution

$$dS \text{ rot } u = \tfrac{1}{2}\int dr\,(u - \nabla\{ur\} + u) - \tfrac{1}{2}u_0\int dr.$$

Now when integrated round dS the perfect differentials dr and $dr\,\nabla\{ur\}$ will vanish. Hence

$$dS \text{ rot } u = \int dr\,u = \int ds\,u,$$

the integral being taken round the margin of dS. Summing over all the elements dS the line integrals along the internal arcs cut out and we obtain Stokes' theorem in its usual form.

19. If in Stokes' theorem we replace u by $B\phi$ as before, we obtain

$$\int\{ds\,.\,B\phi\} = \int\{dS\,[\nabla\,.\,B\phi]\}$$
$$= \int\{dS\,([\nabla_1\,.\,B\phi] + [\nabla_2\,.\,B\phi])\}.$$

Thus $\quad\int\{ds\,B\}\,\phi = \int\{dS\,(\phi\text{ rot }B - [B\nabla]\,\phi)\}$ ....(7).

20. If T be a unit vector along the arc ds whose direction is (l', m', n'), the direction of the normal N to dS being (l, m, n) as before, Stokes' theorem is

Analogue of Stokes' theorem.

$$\int dS\,\{N\text{ rot }u\} = \int ds\,\{Tu\},$$

or $\quad \int dS \begin{vmatrix} l\,, & m\,, & n \\ \dfrac{d}{dx}, & \dfrac{d}{dy}, & \dfrac{d}{dz} \\ u\,, & v\,, & w \end{vmatrix} = \int ds\,(l'u + m'v + n'w).$

Since this is analytically true for all values of u, v, w, we may put $u = i\phi, v = j\phi, w = k\phi$, where ϕ is a scalar function. Then

$$\int dS \begin{vmatrix} i\,, & j\,, & k \\ l\,, & m\,, & n \\ \dfrac{d}{dx}, & \dfrac{d}{dy}, & \dfrac{d}{dz} \end{vmatrix}\phi = \int ds\,(il' + jm' + kn')\,\phi,$$

$$\therefore \int dS\,[N\nabla]\,\phi = \int ds\,T\phi,$$

or $\qquad\qquad \int[dS\,\nabla]\,\phi = \int ds\,\phi$ ....(8).

21. We have

Lemma III.
$$\text{div}\,[\mathbf{AB}] = \{\boldsymbol{\nabla}\,[\mathbf{AB}]\},$$

and we may replace $\boldsymbol{\nabla}$ by $\boldsymbol{\nabla}_1 + \boldsymbol{\nabla}_2$, where $\boldsymbol{\nabla}_1$ operates on $\mathbf{A}$ only and $\boldsymbol{\nabla}_2$ on $\mathbf{B}$ only; further if $\mathbf{C}$ be any vector

$$\mathbf{C}\,[\mathbf{AB}] = \mathbf{B}\,[\mathbf{CA}] = -\mathbf{A}\,[\mathbf{CB}].$$

Hence
$$\text{div}\,[\mathbf{AB}] = \mathbf{B}\,[\boldsymbol{\nabla}_1\mathbf{A}] - \mathbf{A}\,[\boldsymbol{\nabla}_2\mathbf{B}]$$
$$= \mathbf{B}\,\text{rot}\,\mathbf{A} - \mathbf{A}\,\text{rot}\,\mathbf{B} \quad \dots\dots\dots\dots(9).$$

Lemma IV.   **22.** In a similar manner

$$\text{rot}\,[\mathbf{AB}] = [\boldsymbol{\nabla}\,[\mathbf{AB}]]$$
$$= [\boldsymbol{\nabla}_1\,[\mathbf{AB}]] + [\boldsymbol{\nabla}_2\,[\mathbf{AB}]]$$
$$= (\{\mathbf{B}\boldsymbol{\nabla}_1\}\,\mathbf{A} - \mathbf{B}\,\{\boldsymbol{\nabla}_1\mathbf{A}\}) + (\mathbf{A}\,\{\boldsymbol{\nabla}_2\mathbf{B}\} - \{\mathbf{A}\boldsymbol{\nabla}_2\}\,\mathbf{B})$$
$$= \mathbf{B}\boldsymbol{\nabla}.\mathbf{A} - \mathbf{B}\,\text{div}\,\mathbf{A} + \mathbf{A}\,\text{div}\,\mathbf{B} - \mathbf{A}\boldsymbol{\nabla}.\mathbf{B}\dots\dots(10).$$

23. We have seen that the operator $\boldsymbol{\nabla}$, whether operating
Operator ∇^2.
on a scalar or vector quantity, has a meaning independent of the axes of reference; hence the operator $\{\boldsymbol{\nabla}\boldsymbol{\nabla}\}$ must also be independent of the axes. We may obtain the meaning in the following manner:—

If ϕ be the value of a function at a point (x, y, z) whose vector from the origin is $\mathbf{r}$, then at a neighbouring point $\mathbf{r}+\boldsymbol{\rho}$, where $\boldsymbol{\rho}$ or (ξ, η, ζ) is small, the value of the function will, by Taylor's theorem, be

$$\phi + \left(\xi\frac{d\phi}{dx} + \eta\frac{d\phi}{dy} + \zeta\frac{d\phi}{dz}\right)$$
$$+ \tfrac{1}{2}\left\{\xi^2\frac{d^2\phi}{dx^2} + \eta^2\frac{d^2\phi}{dy^2} + \zeta^2\frac{d^2\phi}{dz^2}\right.$$
$$\left. + 2\eta\zeta\frac{d^2\phi}{dy\,dz} + 2\zeta\xi\frac{d^2\phi}{dz\,dx} + 2\xi\eta\frac{d^2\phi}{dx\,dy}\right\}$$
$$+ \text{higher powers of } \xi, \eta, \zeta.$$

Now the mean values of ξ^2, η^2 and ζ^2 over the surface of a sphere of small radius ρ are each equal to $\tfrac{1}{3}(\xi^2+\eta^2+\zeta^2)$, or $\tfrac{1}{3}\rho^2$; while the mean of all terms including odd powers of ξ,

η or ζ is zero. Thus the mean value of the function over the surface is

$$\phi + \frac{\rho^2}{6} \nabla^2\phi + \text{fourth and higher powers of } \rho :$$

and we find

$$\nabla^2\phi = \frac{6}{\rho^2} \{\text{the excess of the mean value of } \phi \text{ over a spherical surface of small radius } \rho \text{ above its value at the centre}\}.$$

We may thus call $\nabla^2\phi$ the 'dispersion' of ϕ.

CHAPTER II.

APPLICATIONS OF VECTORIAL METHODS TO MAGNETOSTATICS.

24. THE potential at a point (x', y', z') due to a magnetic

Potential of a magnetic doublet.

pole of strength μ at (x, y, z) is μ/r, or μp, where
$$r = p^{-1} = \{(x' - x)^2 + (y' - y)^2 + (z' - z)^2\}^{\frac{1}{2}}.$$

Let us consider a magnetic doublet consisting of poles $-\mu$ and $+\mu$ at P, P' respectively; and let the length PP' be ρ and its direction $\mathbf{D}$, the scalar magnitude of this last vector being unity. If p and p' be the reciprocals of the distances of P and P' from the point (x', y', z'), the magnetic potential there due to the doublet will be $\mu p' - \mu p$ or $\mu(p' - p)$. Now p' differs from p in that it is estimated at a point distant ρ from P in the direction $\mathbf{D}$. Hence $p' - p$ is equal, when ρ is indefinitely small, to $\rho \times$ (rate of change of p in the direction $\mathbf{D}$) or $\rho \cdot \mathbf{D}\nabla \cdot p$ by § 11. Hence $\Omega = \mu\rho \cdot \mathbf{D}\nabla \cdot p$ or $M \cdot \mathbf{D}\nabla \cdot p$, if while ρ diminishes indefinitely μ increases indefinitely in such a manner that $\mu\rho$ remains equal to M. If the vector $M\mathbf{D}$ be denoted by $\mathbf{M}$, so that $\mathbf{M}$ has the moment and direction of the magnetic doublet, this may be put into the form $\mathbf{M}\nabla \cdot p$.

In an exactly similar manner it may be shown that the potential energy of the doublet in a field whose potential is Ω is $\mu(\Omega' - \Omega)$ or $\mathbf{M}\nabla \cdot \Omega$.

25. This analysis shows that a magnetic moment obeys

A magnetic doublet is a vector.

the laws of a vector; and the truth of this is obvious from the fact that we can introduce equal and opposite poles $+\mu$ and $-\mu$ at the ends U,

V of the rectangular components $l\rho$, $m\rho$, $n\rho$, and thereby

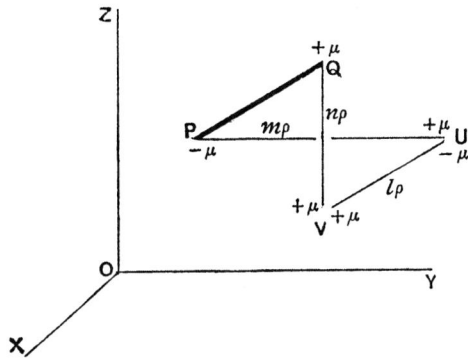

completely replace our doublet PQ of moment M by doublets PU, UV, VQ of moments lM, mM, nM respectively.

26. Let us consider a magnetised body of which I is the

Potential of a magnetised body. intensity of magnetisation. Due to an element of volume δv at (x, y, z) of moment $I\delta v$ the potential at a point (x', y', z') outside the body is $\delta v . I\nabla . p$. Due then to the whole body

$$\Omega = \int dv . I\nabla . p \quad\quad\quad\quad\dots\dots\dots\dots(11)$$

$$= -\int dS . NI . p - \int dv\, p\, \text{div } I, \text{ by (3)},$$

$$= \int dS\, \sigma p + \int dv\, \rho p \quad\quad\dots\dots\dots\dots(12),$$

where $\quad\quad \sigma = -NI, \quad \rho = -\text{div } I \quad\quad\dots\dots\dots\dots(13)$.

Thus the magnetic potential is the same as that due to a surface density, equal to the normal outward component of magnetisation, and a volume density which is *minus* the divergency of the magnetisation.

27. The potential (11) above found is, strictly speaking, applicable only at external points, for $I\nabla . p$ is infinite at internal points. The expression (12) is however finite at internal as well as external points if σ, ρ are finite. Now the potential inside a magnet, regarded as made up of doublets, will change with extreme rapidity as we pass from doublet to doublet, and we may suppose that the number of doublets in unit volume is very large. Thus the potential at (x, y, z) is in reality indefinite unless account is taken of the distribution of doublets in its immediate neighbourhood, and we cannot specify it in terms of I alone. But we may for convenience define the value of Ω given by (12) as the magnetic potential within the body; and in that case, from the ordinary properties of the potential,

$$\left.\begin{aligned} \nabla^2\Omega + 4\pi\rho &= 0 \\ (\mathbf{N}\nabla.\Omega)_1^2 + 4\pi\sigma &= 0 \end{aligned}\right\} \quad\ldots\ldots\ldots\ldots(14).$$

Now if the magnetic force $-\nabla\Omega$ be denoted by $\mathbf{H}$ these equations become

$$\left.\begin{aligned} -\operatorname{div}\mathbf{H} - 4\pi\operatorname{div}\mathbf{I} &= 0 \\ -(\mathbf{NH})_1^2 - 4\pi(\mathbf{NI})_1^2 &= 0 \end{aligned}\right\},$$

where in the second equation we have written $(\mathbf{NI})_1^2$ or $N_1 I_1 + N_2 I_2$ in order to include cases of contact between two magnetised bodies.

28. Thus if we introduce a new vector $\mathbf{B}$ defined by the equation $\mathbf{B} = \mathbf{H} + 4\pi\mathbf{I}$, we have

Magnetic induction.

$$\left.\begin{aligned} \operatorname{div}\mathbf{B} &= 0 \\ (\mathbf{NB})_1^2 &= 0 \end{aligned}\right\} \quad\ldots\ldots\ldots\ldots\ldots(15).$$

This quantity is called by Maxwell the magnetic induction, and its distribution, being like that of the velocity of an incompressible fluid, may be called solenoidal.

29. Since div $\mathbf{B}$ vanishes we may, without loss of generality, suppose that the rectangular components of $\mathbf{B}$ are

The vector potential.

$$\frac{dH}{dy} - \frac{dG}{dz}, \quad \frac{dF}{dz} - \frac{dH}{dx}, \quad \frac{dG}{dx} - \frac{dF}{dy},$$

i.e. $\mathbf{B} = \operatorname{curl}\mathbf{A}$, where $\mathbf{A} \equiv (F, G, H)$.

Then Stokes' theorem gives over any surface

$$\int\{\mathbf{dS}\,\mathbf{B}\} = \int\{\mathbf{ds}\,\mathbf{A}\},$$

and in free space, as $\mathbf{B} = \mathbf{H}$, the surface integral of normal force over any area is equal to the line integral of the tangential component of $\mathbf{A}$ round its margin. The name given to $\mathbf{A}$ by Maxwell is the vector potential.

30. Owing to the presence of a magnetic doublet of moment **M** at (x, y, z) we shall have at (x', y', z')

$$\Omega = \mathbf{M}\nabla \cdot p = -\mathbf{M}\nabla' \cdot p,$$

where $\nabla' \equiv \mathbf{i}\dfrac{d}{dx'} + \mathbf{j}\dfrac{d}{dy'} + \mathbf{k}\dfrac{d}{dz'}$, and as p is a function of $x' - x$,

$y' - y$ and $z' - z$, we have $\dfrac{dp}{dx'} = -\dfrac{dp}{dx}$, $\dfrac{dp}{dy'} = -\dfrac{dp}{dy}$, $\dfrac{dp}{dz'} = -\dfrac{dp}{dz}$.

Thus
$$\mathbf{H} = \nabla' \{\mathbf{M}\nabla'\} p$$
$$= (\nabla' \cdot \mathbf{M}\nabla' - \mathbf{M}\cdot\nabla'^2) p,$$

for, as p satisfies Laplace's equation, $\nabla'^2 p = 0$,

$$= [\nabla'[\nabla'\mathbf{M}]]p, \text{ by Lemma I, paragraph 9,}$$
$$= -\operatorname{rot}'[\mathbf{M}\nabla']p, \text{ where } \operatorname{rot}'\mathbf{C} \equiv [\nabla'\mathbf{C}].$$

Hence we may take, as due to the doublet,

$$A = -[\mathbf{M}\nabla']p = [\mathbf{M}\nabla]p.$$

31. Considering a body of which I is the magnetisation at (x, y, z), the vector potential at (x', y', z') will be

$$\int dv\,[\mathbf{I}\nabla]\,p;$$

or, by Green's theorem,

$$-\int dS\,[\mathbf{I N}]\,p - \int dv\,[\mathbf{I}\nabla_1]\,p,$$

where in the second term ∇_1 must be regarded as operating on I but not on p.

Thus
$$A = \int dS\,[\mathbf{N I}]\,p + \int dv\,[\nabla_1 \mathbf{I}]\,p$$
$$= \int dS\,[\mathbf{N I}]\,p + \int dv\,(\operatorname{rot}\mathbf{I})\,p,$$

and the vector potential may be regarded as due to a surface density $[\mathbf{N I}]$ and a volume density rot I.

32. Let us express by these methods the mutual energy of two simple magnetic shells of moments ϕ, ϕ' per unit area.

Mutual energy of two shells.

We have seen that the vector potential at (x', y', z') due to a magnetic particle at (x, y, z) is

$$A = -[\mathbf{M}\nabla']p = [\mathbf{M}\nabla]p.$$

Now the first shell may be regarded as made up of elements of area dS of which the magnetic moment is ϕdS and direction $\phi N\, dS$: thus its vector potential at $(x',\, y',\, z')$ will be

$$A = \phi \int dS\, [\mathbf{N} \boldsymbol{\nabla}]\, p$$
$$= \phi \int [\mathrm{d}\mathbf{S}\; \boldsymbol{\nabla}]\, p,$$

and this, by the analogue of Stokes' theorem, is equal to

$$\phi \int \mathrm{d}\mathbf{s}\; p.$$

Now, by § 24, for a particle of moment $\mathbf{M}'$ at $(x',\, y',\, z')$ the energy of position W is equal to $\mathbf{M}'\boldsymbol{\nabla}'.\,\Omega$. Thus for the two shells, regarding the second as made up of doublets $\phi' d\mathbf{S}'$,

$$W = \phi' \int dS'\; \{\mathbf{N}'\,\boldsymbol{\nabla}'\}\; \Omega$$
$$= - \phi' \int \{\mathrm{d}\mathbf{S}'\; \mathbf{H}'\}$$
$$= - \phi' \int \{\mathrm{d}\mathbf{S}'\, \mathrm{rot}'\, \mathbf{A}\}$$
$$= - \phi' \int \mathrm{d}\mathbf{s}'\, \mathbf{A}, \text{ by Stokes' theorem,}$$
$$= - \phi\phi' \iint \mathrm{d}\mathbf{s}\; \mathrm{d}\mathbf{s}'\, p$$
$$= - \phi\phi' \iint ds\; ds'\, \cos \epsilon / r \; \dots\dots\dots\dots(16),$$

where ϵ is the angle between the directions of $ds,\, ds'$.

33. When the magnetising force is extremely small the
Induced magnetisation. induced temporary magnetisation $\mathbf{I}_t$ is proportional to the magnetic force $\mathbf{H}$ and is equal to $k\mathbf{H}$, where k is the susceptibility. In order that the analysis may include cases both of temporary magnetisation $\mathbf{I}_t$ and permanent magnetisation $\mathbf{I}_p$, we shall suppose that both may exist together and thus assume that the total magnetisation $\mathbf{I}$ is equal to $\mathbf{I}_t + \mathbf{I}_p$ or $k\mathbf{H} + \mathbf{I}_p$. Thus

$$\mathbf{B} = \mathbf{H} + 4\pi\mathbf{I} = \mathbf{H} + 4\pi\, (k\mathbf{H} + \mathbf{I}_p)$$
$$= \mu\mathbf{H} + 4\pi\mathbf{I}_p, \text{ where } \mu = 1 + 4\pi k \; \dots\dots(17).$$

Thus the conditions (15) obtained in § 28, i.e.

$$\mathrm{div}\, \mathbf{B} = 0, \quad (\mathbf{NB})_1^2 = 0,$$

give $\mathrm{div}\, \mu\mathbf{H} + 4\pi\, \mathrm{div}\, \mathbf{I}_p = 0, \quad \{\mathbf{N}\, (\mu\mathbf{H} + 4\pi\mathbf{I}_p)\}_1^2 = 0.$

We now replace $\mathbf{H}$ by $-\boldsymbol{\nabla}\Omega$ and denote the permanent magnetic densities, corresponding to those of (13) in § 25, by $\rho_p,\, \sigma_p$, i.e.

$$\sigma_p = - \{\mathbf{N}\mathbf{I}_p\}_1^2, \quad \rho_p = - \mathrm{div}\, \mathbf{I}_p\, ;$$

then $\operatorname{div} \mu \nabla \Omega + 4\pi \rho_p = 0, \quad \{N \mu \nabla \Omega\}_1^2 + 4\pi \sigma_p = 0$;

or $\nabla_\mu^2 \Omega + 4\pi \rho_p = 0, \quad \left(\mu \dfrac{d\Omega}{dn} \right)_1^2 + 4\pi \sigma_p = 0 \ \ldots\ldots (18),$

where $\nabla_\mu^2 \Omega$ denotes $\dfrac{d}{dx} \left(\mu \dfrac{d\Omega}{dx} \right) + \dfrac{d}{dy} \left(\mu \dfrac{d\Omega}{dy} \right) + \dfrac{d}{dz} \left(\mu \dfrac{d\Omega}{dz} \right)$ and dn is an element of the normal drawn *into* the region of the corresponding potential.

34. The potential energy of a magnetostatic field may be obtained by considering the work done in gradually and proportionally increasing the strength of all the permanent magnets from zero to their final value; during this process any iron capable of temporary magnetisation must remain in its final position, its magnetisation at any time being determined by the field due to the permanent magnets. At a time when all the permanent magnetisation is of n times its final strength the value of the potential and force at any point will be n times the final value and, as in the case of an electrostatic field, the work done in increasing n from n to $n+\delta n$ will be $\Sigma n \delta n m_p \Omega$, where m_p is a representative permanent magnetic pole. Thus the work done in creating the system will be

The potential energy of a magneto-static pole.

$$W = \tfrac{1}{2} \Sigma m_p \Omega = \tfrac{1}{2} \Sigma \{M_p \nabla\} \, \Omega,$$

where M_p is the moment of a representative permanent magnetic doublet,

$$= -\tfrac{1}{2} \Sigma \{M_p H\}$$

$$= -\tfrac{1}{2} \int dv \{I_p H\}$$

$$= -\frac{1}{8\pi} \int dv \{(B - \mu H) H\}$$

$$= +\frac{1}{8\pi} \int dv \mu H^2 + \frac{1}{8\pi} \int dv \, B \nabla . \Omega$$

$$= \frac{1}{8\pi} \int dv \mu H^2 - \frac{1}{8\pi} \int dS \{NB\}_1^2 \Omega - \frac{1}{8\pi} \int dv \Omega \operatorname{div} B,$$

by (3) of § 14.

Now, by (15) of § 28, $\{NB\}_1^2 = 0$, $\operatorname{div} B = 0$: hence

$$W = \frac{1}{8\pi} \int dv \mu H^2 = \frac{1}{8\pi} \int dv \mu H^2 \ldots\ldots\ldots\ldots (19).$$

CHAPTER III.

35. In his papers and his classical Treatise on Electricity and Magnetism Maxwell gave a number of different interpretations of the processes at work, and the interest of these caused nearly as much importance to be attached to them as to the final equations to which they led. It was Hertz who pointed out that however Maxwell's equations might be interpreted it was they which in effect constituted Maxwell's theory, and he put them into an extremely convenient form.

In the electrostatic-electromagnetic units adopted by Hertz the energy of the field per unit volume is taken, when the media are stationary, as $\frac{1}{8\pi}(K\mathsf{E}^2 + \mu\mathsf{H}^2)$, where the units are such that for free space $K = 1$ and $\mu = 1$, and E, H stand for the electric and magnetic forces.

We adopt the following further symbols:

$\mathsf{D} = K\mathsf{E}$ = the electric polarisation
    = Maxwell's displacement multiplied by 4π,
C = the conduction current,
B = the magnetic polarisation
    = Maxwell's magnetic induction = $\mu\mathsf{H} + 4\pi\mathsf{I}_p$*.

Then Hertz's equations are

$$\left. \begin{array}{c} \dfrac{d\mathsf{D}}{dt} + 4\pi\mathsf{C} = V \operatorname{rot} \mathsf{H} \\[2mm] \dfrac{d\mathsf{B}}{dt} = - V \operatorname{rot} \mathsf{E} \end{array} \right\} \quad\ldots\ldots\ldots\ldots..(20).$$

* Hertz does not explicitly discuss the case of permanent magnetisation.

36. It follows that over any area

$$\int \left\{ d\mathbf{S} \left(\frac{d\mathbf{D}}{dt} + 4\pi\mathbf{C} \right) = V \int \{ d\mathbf{S} \text{ rot } \mathbf{H} \} \right.$$

$$= V \int d\mathbf{s} \, \mathbf{H} \, ;$$

thus the rate of increase of the surface integral of electric polarisation over any area, together with 4π times the conduction current through it, is equal to the line integral of $\mathbf{H}$ round it. Similarly from the second equation it follows that the rate of diminution of the magnetic polarisation through any circuit is equal to the line integral of $\mathbf{E}$ round it. Thus the equations (20) express Maxwell's fundamental relations.

37. Also, taking the divergence of the former equation of (20),

$$\text{div} \left(\frac{d\mathbf{D}}{dt} + 4\pi\mathbf{C} \right) = 0 \, ;$$

but, by the definition of the conduction current, div $\mathbf{C}$ is the rate at which charge is conveyed away per unit volume, and must be equal to $-\dfrac{d\rho}{dt}$, where ρ is the electric density. Hence

$$\text{div} \frac{d\mathbf{D}}{dt} - 4\pi \frac{d\rho}{dt} = 0$$

at all points, and integrating with reference to the time,

$$\text{div } \mathbf{D} = 4\pi\rho,$$

the constant of integration vanishing, since $\rho = 0$ at all points if $\mathbf{D} = 0$ at all points.

Similarly $\qquad\qquad$ div $\mathbf{B} = 0.$

38. In electrostatic fields $\dfrac{d\mathbf{B}}{dt} = 0$, for there is no time change of any variable. Hence rot $\mathbf{E} = 0$, and we may take $\mathbf{E} = -\nabla\phi$, where ϕ is a function given by

$$4\pi\rho = \text{div } \mathbf{D} = -\text{div} \left(K\nabla\phi \right) = -\nabla_K^2 \phi,$$

$$4\pi\sigma = \{\mathbf{N}\mathbf{D}\}_1^2 = -\left(K \frac{d\phi}{dn} \right)_1^2 .$$

$$2\text{---}2$$

Similarly for a magnetostatic field $\mathbf{H} = -\nabla\Omega$, where Ω is given by

$$0 = \operatorname{div}\mathbf{B}, \quad 0 = \{\mathbf{NB}\}_1^2,$$

as in § 33.   Thus

$$\nabla_\mu^2\Omega + 4\pi\rho_p = 0 \left.\begin{array}{l} \\ \\ \end{array}\right\},$$
$$\left(\mu\frac{d\Omega}{dn}\right)_1^2 + 4\pi\sigma_p = 0$$

where ρ_p, σ_p are the volume and surface densities of permanent magnetism.

39. A surface at which there is a discontinuity between
Surface the physical conditions on the two sides should
conditions. be regarded as the limit of a thin layer of
continuous transition when the thickness of the layer is indefinitely diminished.   Now the values of $\dfrac{d\mathbf{D}}{dt} + 4\pi\mathbf{C}$, and of $\dfrac{d\mathbf{B}}{dt}$, are finite on each side of the layer, and so may be regarded as finite within it also: hence the values of rot $\mathbf{H}$ and rot $\mathbf{E}$ will be finite in the layer.   But if the axis of Z be taken in the direction of the normal to the bounding surface at any point the first of the three rectangular components of rot $\mathbf{H}$ will be $\dfrac{dN}{dy} - \dfrac{dM}{dz}$, where $\mathbf{H} = (L, M, N)$.   Now as $\dfrac{dN}{dy}$ is finite on each side of the layer it will be finite within it: hence $\dfrac{dM}{dz}$ will also be finite in the layer, and $\displaystyle\int_1^2 dz\,\dfrac{dM}{dz}$ integrated through the layer will be of the same order of small quantities as the thickness of the layer.   Hence when the thickness is indefinitely diminished the values of M on the two sides will be the same.   In a similar manner the values of L, X, Y may be shown to be the same on the two sides.   Thus the tangential electric and magnetic forces must be continuous across the surface.

40. It follows from the consideration of adjacent points in the first medium, and points opposite to them in the second medium, that the differential coefficients of L, M, X or Y with respect to x or y will be the same on the two sides*.   Hence in the third of the three Cartesian equations of the former of (20), i.e.

$$K \frac{dZ}{dt} + 4\pi r = V \left(\frac{dM}{dx} - \frac{dL}{dy} \right),$$

the terms on the right side are continuous.   Accordingly the left side must have the same value on the two sides, i.e. the total normal electric flow is continuous.

Similarly the normal magnetic flow is continuous.

It must be noted that as the two boundary conditions of this section are derived from the equations of the field, and these equations are satisfied throughout each medium, the two boundary conditions are satisfied when the conditions of § 39 are satisfied.   Hence the independent boundary conditions reduce to the continuity of the tangential components of E and H.

41.   It follows from $Y_1 - Y_2 = 0$, $Z_1 - Z_2 = 0$ that $(\mathsf{E}_1 - \mathsf{E}_2)$ must be normal in direction and hence that $[\mathsf{N}\,(\mathsf{E}_1 - \mathsf{E}_2)]$ must be zero.   Thus the surface conditions may be put into the form

$$[\mathsf{NE}]_1^2 = 0, \quad [\mathsf{NH}]_1^2 = 0 \quad\ldots\ldots\ldots\ldots\ldots(21).$$

* If we take points P, P' in the first medium such that PP' is parallel to OX, and if Q, Q' be the points closest to them in the second medium, such that the lengths PQ and $P'Q'$ are of the second order of small quantities, then in the first medium $\frac{dM}{dx}$ is the limit of $\frac{M_{P'} - M_P}{PP'}$; and as $M_P = M_Q$ and $M_{P'} = M_{Q'}$ this is the same in the limit as $\frac{M_Q - M_Q}{QQ'}$ or $\frac{dM}{dx}$ in the second medium.

CHAPTER IV.

HERTZ'S EQUATIONS FOR MOVING MEDIA.

42. WE have next to consider the case in which the material media in which electromagnetic processes are at work are in motion, and we shall suppose that the velocity at any point is $\mathbf{u}$, a continuous function of the coordinates.

The most natural extension of the two fundamental laws of Maxwell is to suppose that they apply to a circuit moving with the velocity $\mathbf{u}$ of the medium. Now the rate of change of the surface integral of any vector $\mathbf{R}$ is made up of two parts, one due to the change in $\mathbf{R}$ and the other to the motion of the surface. The former part is $\int d\mathbf{S}\, \dfrac{d\mathbf{R}}{dt}$; the latter may be obtained by considering the cylindrical element of volume δv whose ends δS, $\delta S'$ are formed by the area δS in its positions at the time t and the time $t + \delta t$. The total surface integral of the component of $\mathbf{R}$ along the inward normals to the element of volume is by Green's theorem equal to $-\delta v\,\mathrm{div}\,\mathbf{R}$, or $-\{\boldsymbol{\delta}\mathbf{S}\,\mathbf{u}\,\delta t\}\,\mathrm{div}\,\mathbf{R}$.

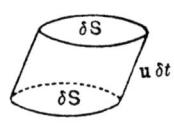

Now the contribution to the surface integral from the two faces δS, $\delta S'$ will be

$$\mathbf{R}\,\boldsymbol{\delta}\mathbf{S} - \mathbf{R}'\,\boldsymbol{\delta}\mathbf{S}',$$

where $\mathbf{R}'$ is the value of $\mathbf{R}$ at the face $\boldsymbol{\delta}\mathbf{S}'$. Also the tubular surface may be regarded as made up of parallelograms of which adjacent sides are elements of arc $\boldsymbol{\delta}\mathbf{s}$ (bounding $\boldsymbol{\delta}\mathbf{S}$) and lines u δt: thus by § 9 the contribution to the surface integral from the tubular surface will be

$$-\int\{R[\boldsymbol{\delta}\mathbf{s}\,\mathbf{u}\,\delta t]\} \quad \text{or} \quad \delta t \int \boldsymbol{\delta}\mathbf{s}\,[R\mathbf{u}]$$

$$\text{or} \quad \delta t\,\{\boldsymbol{\delta}\mathbf{S}\,\text{rot}\,[R\mathbf{u}]\}.$$

Hence Green's theorem gives

$$-\{\mathbf{u}\,\boldsymbol{\delta}\mathbf{S}\}\,\delta t\,\text{div}\,\mathbf{R} = \mathbf{R}\,\boldsymbol{\delta}\mathbf{S} - \mathbf{R}'\,\boldsymbol{\delta}\mathbf{S}' + \delta t\,\{\boldsymbol{\delta}\mathbf{S}\,\text{rot}\,[R\mathbf{u}]\};$$

thus $\mathbf{R}'\,\boldsymbol{\delta}\mathbf{S}' - \mathbf{R}\,\boldsymbol{\delta}\mathbf{S} = \delta t\,\{\boldsymbol{\delta}\mathbf{S}\,(\mathbf{u}\,\text{div}\,\mathbf{R} + \text{rot}\,[R\mathbf{u}])\}.$

Accordingly the rate of change of $\mathbf{R}\,\boldsymbol{\delta}\mathbf{S}$ due solely to the motion is $\dot{\mathbf{R}}_1\,\boldsymbol{\delta}\mathbf{S}$, where

$$\dot{\mathbf{R}}_1\,\boldsymbol{\delta}\mathbf{S} = \{\boldsymbol{\delta}\mathbf{S}\,(\mathbf{u}\,\text{div}\,\mathbf{R} + \text{rot}\,[R\mathbf{u}])\},$$

and so $\dot{\mathbf{R}}_1 = \mathbf{u}\,\text{div}\,\mathbf{R} + \text{rot}\,[R\mathbf{u}]$ ....(22).

43. If then we decide to make the assumption that the polarisations D, B are, in spite of the motion, equal to $K\mathbf{E}$, $\mu\mathbf{H} + 4\pi\mathbf{I}_p$, where E, H are the electric and magnetic forces acting in the moving media, we shall have, instead of the equation

$$\int\left\{d\mathbf{S}\left(\frac{d\mathbf{D}}{dt} + 4\pi\mathbf{C}\right)\right\} = V\int\{d\mathbf{S}\,\text{rot}\,\mathbf{H}\},$$

the modified equation

$$\int\left\{d\mathbf{S}\left(\frac{d\mathbf{D}}{dt} + \mathbf{u}\,\text{div}\,\mathbf{D} + \text{rot}\,[D\mathbf{u}] + 4\pi\mathbf{C}\right)\right\} = V\int\{d\mathbf{S}\,\text{rot}\,\mathbf{H}\},$$

and since this is true for all circuits $\boldsymbol{\delta}\mathbf{S}$ we shall have

$$\frac{d\mathbf{D}}{dt} + \mathbf{u}\,\text{div}\,\mathbf{D} + \text{rot}\,[D\mathbf{u}] + 4\pi\mathbf{C} = V\,\text{rot}\,\mathbf{H} \quad......(23).$$

Similarly the second fundamental equation becomes

$$\frac{d\mathbf{B}}{dt} + \mathbf{u}\,\text{div}\,\mathbf{B} + \text{rot}\,[B\mathbf{u}] = -\,V\,\text{rot}\,\mathbf{E} \quad.........(24).$$

On expanding rot [Du] the former equation becomes

$$\frac{d\mathbf{D}}{dt} + \mathbf{u}\nabla\,.\,\mathbf{D} + \mathbf{D}\,\text{div}\,\mathbf{u} - \mathbf{D}\nabla\,.\,\mathbf{u} + 4\pi\mathbf{C} = V\,\text{rot}\,\mathbf{H},$$

or, if the time-rate of change in the value of a function at a point moving with the medium be denoted by $\dfrac{d}{dt'}$, so that

$$\frac{d\mathbf{D}}{dt} + \mathbf{u}\nabla.\mathbf{D} = \frac{d\mathbf{D}}{dt'},$$

$$\frac{d\mathbf{D}}{dt'} + \mathbf{D}\operatorname{div}\mathbf{u} - \mathbf{D}\nabla.\mathbf{u} + 4\pi\mathbf{C} = V\operatorname{rot}\mathbf{H} \ \ldots\ldots(25).$$

Similarly the second fundamental equation becomes

$$\frac{d\mathbf{B}}{dt'} + \mathbf{B}\operatorname{div}\mathbf{u} - \mathbf{B}\nabla.\mathbf{u} = -V\operatorname{rot}\mathbf{E} \ \ldots\ldots(26).$$

These are Hertz's equations for moving media.

44. At a point in the boundary between two media,
Boundary conditions. regarded as the limit of a region of very rapid transition from one medium to the other, we have

$$\frac{d\mathbf{B}}{dt'} + \mathbf{B}\operatorname{div}\mathbf{u} - \mathbf{B}\nabla.\mathbf{u},$$

and

$$\frac{d\mathbf{D}}{dt'} + \mathbf{D}\operatorname{div}\mathbf{u} - \mathbf{D}\nabla.\mathbf{u} + 4\pi\mathbf{C},$$

are finite*, provided that $\mathbf{u}$ and its differential coefficients are finite, i.e. provided that there is no discontinuity in the velocity of the two media at the interface. In that case, by (25) and (26), rot $\mathbf{E}$ and rot $\mathbf{H}$ will be finite in the region of transition, and as in §§ 39, 41 it follows that

$$[\mathbf{NE}]_1^2 = 0, \quad [\mathbf{NH}]_1^2 = 0.$$

45. In order to decide definitely whether the Hertzian
Blondlot's experimental test of Hertz's theory for moving media. theory for moving media is in accordance with facts, Blondlot took two parallel plates, say $z = \pm a$, and made a field of magnetic force L, parallel to OX, between them. He then sent a current

---

* It might at first sight appear legitimate to suppose that in (23)

$$\frac{d\mathbf{D}}{dt} + \mathbf{u}\operatorname{div}\mathbf{D} + 4\pi\mathbf{C}$$

is finite at the interface; and to deduce that the tangential components of $\mathbf{H} - \dfrac{1}{V}[\mathbf{Du}]$ were continuous. This is not however justifiable, for $\dfrac{d}{dt}$ gives the time-change at a stationary point, and while $\dfrac{dK}{dt'}$ is finite $\dfrac{dK}{dt}$ is infinite unless the value of K on the two sides is the same or the velocity $\mathbf{u}$ is parallel to the interface.

of air with velocity v parallel to OY between the plates. According to Hertz's theory this would create opposite charges on the two plates and hence on reversing L or v there would be a current along a wire joining the plates: this was not found and the theory must therefore be incorrect.

In order to obtain the charges we shall suppose that L, v are constant and that the conditions have become steady so that $\frac{d}{dt} = 0$; further the charges on the infinite plates being constant, there will be no current in the wire. As the air is uncharged $\rho = 0$, and in equation (24) $\frac{d\mathbf{B}}{dt} = 0$ both in the air and the metal, $\therefore$ rot $[\mathbf{Bu}] = -V$ rot $\mathbf{E}$ in both media.

$$\therefore \text{ rot }\left(\mathbf{E} + \frac{1}{V}[\mathbf{Bu}]\right) = 0 \quad \ldots\ldots\ldots\ldots(27).$$

Hence we may put in the air,

$$\mathbf{E} + \frac{1}{V}[\mathbf{Bu}] = -\nabla\phi\ldots\ldots\ldots\ldots(27'),$$

where ϕ is given by the condition $\rho = 0$, or

$$\text{div } \mathbf{D} = 4\pi\rho = 0.$$

Now from (27')

$$X = -\frac{d\phi}{dx}, \quad Y = -\frac{d\phi}{dy}, \quad Z = -\frac{d\phi}{dz} - \frac{vL}{V}.$$

Thus, as div $\mathbf{D} = 0$, $\quad -\nabla^2\phi = 0$,

and we take as the appropriate solution

$$\phi = A + Bx + Cy + Dz.$$

Now as the velocity v is parallel to the plates we have $\frac{d\mathbf{D}}{dt} + \mathbf{u}$ div $\mathbf{D}$, and $\frac{d\mathbf{B}}{dt} + \mathbf{u}$ div $\mathbf{B}$ finite in both media, and so finite in the interface*. Thus rot $\left(\mathbf{E} + \frac{1}{V}[\mathbf{Bu}]\right)$ will be finite, and it follows at once that X, Y will be continuous across the surface of the metal.

* See the footnote of the previous page.

But as there is no current in the plates when the conditions are steady X, Y will there vanish.   Hence $B = 0$, $C = 0$ and

$$\phi = A + Dz.$$

In order to determine D we shall utilise the fact that, from (27), the integral over the surface bounded by any circuit

$$\int dS \text{ rot} \left(\mathsf{E} + \frac{1}{V}[\mathsf{Bu}] \right) = 0$$

or the line integral of $\mathsf{E} + \frac{1}{V}[\mathsf{Bu}]$ round the circuit vanishes. Let the circuit consist of a line PP' parallel to OZ in the air from the first plate to the second, a line thence in the second plate from P' to the end of the wire, thence along the wire to the first plate and thence to the point P to complete the circuit. Along the whole of this circuit, except the straight line PP', $\mathsf{E} = 0$ and $\mathsf{u} = 0$.   Hence along PP' the component parallel to OZ of $\mathsf{E} + \frac{1}{V}[\mathsf{Bu}]$ must vanish.

Hence the line integral of $Z + \frac{vL}{V}$, or $-D + \frac{vL}{V}$, must vanish: and $D = \frac{vL}{V}$.

Thus the value of Z within the metal being zero, the surface charge σ on the plate $z = a$ will be given by

$$\frac{vL}{V} = 4\pi\sigma.$$

Similarly on $z = -a$ the charge will be $-\frac{vL}{4\pi V}$ per unit area. Thus on reversing v or L there will be a current in the wire, and its amount can be calculated for comparison with experiment.

46.   Another test which may be applied to Hertz's theory is that of finding the influence of motion in the medium upon the velocity of light propagated along the direction of motion.

The influence upon the velocity of light.

For a medium moving with uniform velocity u parallel to OX the equations are

$$K\left(\frac{dX}{dt} + u\frac{dX}{dx}\right) = V\left(\frac{dN}{dy} - \frac{dM}{dz}\right),$$

$$K\left(\frac{dY}{dt} + u\frac{dY}{dx}\right) = V\left(\frac{dL}{dz} - \frac{dN}{dx}\right),$$

$$K\left(\frac{dZ}{dt} + u\frac{dZ}{dx}\right) = V\left(\frac{dY}{dx} - \frac{dX}{dy}\right),$$

$$\frac{dL}{dt} + u\frac{dL}{dx} = -V\left(\frac{dZ}{dy} - \frac{dY}{dz}\right),$$

$$\frac{dM}{dt} + u\frac{dM}{dx} = -V\left(\frac{dX}{dz} - \frac{dZ}{dx}\right),$$

$$\frac{dN}{dt} + u\frac{dN}{dx} = -V\left(\frac{dY}{dx} - \frac{dX}{dy}\right).$$

Thus if the velocity of propagation be U, and we take

$$\frac{X}{\xi} = \frac{Y}{\eta} = \frac{Z}{\zeta} = \frac{L}{\lambda} = \frac{M}{\mu} = \frac{N}{\nu} = e^{is(Ut-x)},$$

we find

$$\left.\begin{array}{l} K(U-u)\xi = 0 \\ K(U-u)\eta = +V\nu \\ K(U-u)\zeta = -V\mu \end{array}\right\},$$

$$\left.\begin{array}{l} (U-u)\lambda = 0 \\ (U-u)\mu = -V\zeta \\ (U-u)\nu = +V\eta \end{array}\right\}.$$

Hence ξ and λ vanish, and E and H lie in the wave front: and as $\lambda\xi + \mu\eta + \nu\zeta$ vanishes, E and H are at right angles.

Further, eliminating μ, ζ,

$$K(U-u)^2 = V^2 = KV'^2,$$

if V' be the velocity of light in the medium at rest. Hence

$$U = u \pm V'.$$

47. According to this theory then the velocity u of the medium is superposed on that of the light, a result which is contradicted by experimental determinations: the latter show that the velocity of light along a current of air is only affected by a small fraction of the velocity of the air.

CHAPTER V.

SOME EFFECTS DUE TO THE MOTION OF CHARGED PARTICLES THROUGH A STATIONARY AETHER.

48. BEFORE considering Lorentz's theory, in which all phenomena are interpreted in terms of electrons moving through aether at rest, it is of interest to examine some simple cases of this type in which a complete solution can be effected.

Let us consider the case in which electricity of density ρ per unit volume is moving with velocity u through stationary aether, the only restriction on u being that it shall be a finite and continuous function of the coordinates. If we consider a circuit fixed in the aether the convection current will be ρu and the fundamental laws give

$$\left. \begin{array}{c} \dfrac{d\mathsf{E}}{dt} + 4\pi\rho\mathsf{u} = V \operatorname{rot} \mathsf{H} \\[2mm] \dfrac{d\mathsf{H}}{dt} = - V \operatorname{rot} \mathsf{E} \end{array} \right\} \quad \dots\dots\dots\dots(28).$$

Now, as Maxwell pointed out, the electric force E' acting on a conductor moving with velocity u is not the same as E the force when the conductor is at rest: and as E is the same whether the result of it be to set up a conduction current, to act on a charged particle, or to cause polarisation in a dielectric, we shall suppose that E' is the same whatever be the nature of the effect produced on the moving body. Let us consider a circuit of which u is the velocity at any point. The rate of increase of the surface integral of electric force E will be

$$\int d\mathsf{S} \left(\frac{d\mathsf{E}}{dt} + \mathsf{u} \operatorname{div} \mathsf{E} + \operatorname{rot} [\mathsf{Eu}] \right);$$

and there will be no convection current through the circuit
as the velocity of the charge relative to the circuit is zero.
Thus if we apply the first fundamental law to the moving
circuit we get

$$\int d\mathbf{S} \left(\frac{d\mathbf{E}}{dt} + \mathbf{u} \operatorname{div} \mathbf{E} + \operatorname{rot} [\mathbf{Eu}]\right) = V \int d\mathbf{S} \operatorname{rot} \mathbf{H}',$$

where on the right side we have $\mathbf{H}'$, not $\mathbf{H}$, since it is the
magnetic force on a moving circuit which is considered.

Hence $\qquad \dfrac{d\mathbf{E}}{dt} + 4\pi\rho\mathbf{u} + \operatorname{rot} [\mathbf{Eu}] = V \operatorname{rot} \mathbf{H}'$ ....(28'),

and similarly from the second law

$$\frac{d\mathbf{H}}{dt} + \operatorname{rot} [\mathbf{Hu}] = - V \operatorname{rot} \mathbf{E}' ....(29).$$

On comparing these with (28) we find

$$\operatorname{rot} [\mathbf{Eu}] = V \operatorname{rot} (\mathbf{H}' - \mathbf{H}),$$

and so we take $\qquad \mathbf{H}' = \mathbf{H} + \dfrac{1}{V} [\mathbf{Eu}]$ ....(30).

Similarly from the second equations

$$\mathbf{E}' = \mathbf{E} + \frac{1}{V} [\mathbf{uH}] \qquad ....(31).$$

49. On expanding $\operatorname{rot} [\mathbf{Eu}]$ as

$$\mathbf{E} \operatorname{div} \mathbf{u} - \mathbf{E} \nabla . \mathbf{u} - \mathbf{u} \operatorname{div} \mathbf{E} + \mathbf{u} \nabla . \mathbf{E},$$

**Boundary
conditions.** and replacing $\dfrac{d}{dt} + \mathbf{u}\nabla$ by $\dfrac{d}{dt'}$, we get from (28')

$$\frac{d\mathbf{E}}{dt'} + \mathbf{E} \operatorname{div} \mathbf{u} - \mathbf{E} \nabla . \mathbf{u} = V \operatorname{rot} \mathbf{H}' \qquad ....(32),$$

and similarly from (29)

$$\frac{d\mathbf{H}}{dt'} + \mathbf{H} \operatorname{div} \mathbf{u} - \mathbf{H} \nabla . \mathbf{u} = -V \operatorname{rot} \mathbf{E}'.........(33).$$

Now at a surface which separates two media we shall have,
if $\mathbf{u}$ be continuous at the surface, the left sides of (32) and (33)
finite within the region of rapid transition which replaces the
surface of discontinuity: thus the right sides will be finite, and

the condition at the surface, as in § 44, will be that the tangential components of E', H' shall be the same in the two regions, i.e. that

$$[NE']_1^2 = 0, \quad [NH']_1^2 = 0.$$

50. We shall now consider in slightly greater detail the case in which electric charges of density ρ are moving through empty space with velocity u which is uniform and constant.

Convection with constant velocity.

The equations of the field referred to moving circuits will be, from (32) and (33),

$$\left. \begin{aligned} \frac{d\mathsf{E}}{dt'} &= V \operatorname{rot} \mathsf{H}' \\[2mm] \frac{d\mathsf{H}}{dt'} &= -V \operatorname{rot} \mathsf{E}' \end{aligned} \right\} \quad \dots\dots\dots\dots(34).$$

Also $\qquad \operatorname{div} \mathsf{E} = 4\pi\rho, \quad \operatorname{div} \mathsf{H} = 0 \ \dots\dots\dots\dots(35),$

$$\mathsf{E}' = \mathsf{E} + \frac{1}{V}[u\mathsf{H}], \quad \mathsf{H}' = \mathsf{H} - \frac{1}{V}[u\mathsf{E}] \ \ \dots\dots(36).$$

In (34), since the conditions are steady, there are no changes in quantities estimated at points moving with velocity u, and $\dfrac{d}{dt'} = 0$.

Thus $\operatorname{rot} \mathsf{E}' = 0$, $\operatorname{rot} \mathsf{H}' = 0$; and we may put

$$\mathsf{E}' = -\nabla\Phi, \quad \mathsf{H}' = -\nabla\Omega,$$

where Φ, Ω may be called the electric and magnetic convection potentials.

Also by (36)

$$\operatorname{div} \mathsf{E}' = \operatorname{div} \mathsf{E} + \frac{1}{V}(\mathsf{H} \operatorname{rot} u - u \operatorname{rot} \mathsf{H}) \ \ \dots\dots(37).$$

Now

$$\operatorname{rot} \mathsf{H} = \operatorname{rot}\left(\mathsf{H}' + \frac{1}{V}[u\mathsf{E}]\right)$$

$$= \operatorname{rot} \mathsf{H}' + \frac{1}{V}(u \operatorname{div} \mathsf{E} - u\nabla . \mathsf{E} + \mathsf{E}\nabla . u - \mathsf{E} \operatorname{div} u)$$

$$= \frac{1}{V}(u \operatorname{div} \mathsf{E} - u\nabla . \mathsf{E}),$$

for rot $H' = 0$, and u is independent of the coordinates, being uniform. Thus

$$u \text{ rot } H = \frac{u^2}{V} 4\pi\rho - \frac{1}{V} u\nabla . uE.$$

Also as $u[uH] = 0$,

$$uE = uE' = -u\nabla . \Phi.$$

Thus (37) becomes

$$-\nabla^2 \Phi = 4\pi\rho - \frac{u^2}{V^2} 4\pi\rho - \frac{(u\nabla)^2}{V^2} \Phi,$$

or

$$\left(\nabla^2 - \frac{(u\nabla)^2}{V^2}\right) \Phi + 4\pi\rho \left(1 - \frac{u^2}{V^2}\right) = 0 \ldots\ldots\ldots(38).$$

51. If $u = (u, 0, 0)$ and $1 - \frac{u^2}{V^2} = l^2$,

$$l^2 \frac{d^2\Phi}{dx^2} + \frac{d^2\Phi}{dy^2} + \frac{d^2\Phi}{dz^2} + 4\pi\rho l^2 = 0 \ \ldots\ldots\ldots(39),$$

so that if we put $x = l\xi$, $y = \eta$, $z = \zeta$, we have

$$\frac{d^2\Phi}{d\xi^2} + \frac{d^2\Phi}{d\eta^2} + \frac{d^2\Phi}{d\zeta^2} + 4\pi\rho l^2 = 0,$$

and Φ is the potential at (ξ, η, ζ) due to charges represented by $\rho l^2 d\xi d\eta d\zeta$ within an element of volume $d\xi d\eta d\zeta$.

Thus at (x', y', z')

$$\Phi = \iiint \frac{\rho l \, dx \, dy \, dz}{r_1},$$

where $r_1^2 = (x' - x)^2/l^2 + (y' - y)^2 + (z' - z)^2,$

or

$$\Phi = l \Sigma \frac{e}{r_1},$$

where e is a representative charge in the original system.

In an exactly similar manner, as div $H = 0$,

$$l^2 \frac{d^2\Omega}{dx^2} + \frac{d^2\Omega}{dy^2} + \frac{d^2\Omega}{dz^2} = 0,$$

and as there is now no volume density $\Omega = 0$: hence $H' = 0$, and by (36)

$$H = \frac{1}{V}[uE].$$

Thus when **u** is unrestricted as to direction,

$$- \nabla \Phi = \mathsf{E}' = \mathsf{E} + \frac{1}{V}[\mathsf{uH}]$$

$$= \mathsf{E} + \frac{1}{V^2}[\mathsf{u}\,[\mathsf{uE}]]$$

$$= \mathsf{E} + \frac{1}{V^2}(\mathsf{u}\,.\,\mathsf{uE} - \mathsf{u}^2\,.\,\mathsf{E});$$

$$\therefore \ \left(1 - \frac{\mathsf{u}^2}{V^2}\right)\mathsf{E} + \frac{\mathsf{u}}{V^2}\,.\,\mathsf{uE} = -\,\nabla \Phi.$$

When $\mathsf{u} = (u,\,0,\,0)$ this gives

$$X = -\frac{d\Phi}{dx}, \quad Y = -\frac{1}{l^2}\frac{d\Phi}{dy}, \quad Z = -\frac{1}{l^2}\frac{d\Phi}{dz},$$

and so
$$L = 0, \qquad M = \frac{u}{Vl^2}\frac{d\Phi}{dz}, \qquad N = -\frac{u}{Vl^2}\frac{d\Phi}{dy}.$$

52. Due to a point-charge e moving along OX past the origin $\Phi = le/r$, and if u^2/V^2 be neglected $r_1 = r$, the distance from the origin. Thus the electric forces will be the same as those due to a fixed charge, and, in addition,

$$L = 0, \quad M = -\frac{u}{V}\frac{ez}{r^3}, \quad N = \frac{u}{V}\frac{ey}{r^3}.$$

CHAPTER VI.

THE ELECTRON THEORY OF LORENTZ APPLIED TO STATIONARY MEDIA.

53. WE have seen that the Maxwell-Hertz theory of moving media is contradicted by experience, and it becomes necessary to adopt hypotheses different from those on which that theory was based. It was there supposed that if **E** were the electric force acting upon a circuit moving with the medium, the polarisation of the medium was K**E**; so that if the medium were a greatly rarefied gas the polarisation in the gas would be K times the electric force acting on the moving gas. But in the limit when the gas is evanescent in density the polarisation becomes that of the aether only, and the force tending to produce the polarisation is dependent on the velocity, for, as Maxwell pointed out, the electric force on a moving body exceeds that on a stationary body by $\frac{1}{V}$[u**H**] :

thus the assumption that the polarisation within a moving medium is K times the electric force at a moving point really involves the hypothesis that the aether is carried along with the velocity of the medium. In Lorentz's theory the aether is supposed to be stationary and it is only the electrons (minute particles, either with or without ordinary mass, carrying electric charges) which are supposed to move through it with the velocity of the material medium : the interpretation of electrical phenomena in terms of electrons has received very strong confirmation from the facts of electrolysis, the discharge of gases, kathode and Röntgen rays, radio-activity, electrical conductivity and various optical phenomena. Further as the distributions of

electrons which are required to explain electric and magnetic
polarisation, and conduction currents, are probably very com-
plicated, it becomes desirable for the sake of simplicity to
consider separately the explanation of each of these phenomena;
when they occur simultaneously we can suppose that the effect
may be obtained by adding their separate effects.

54. In the Hertzian electrostatic-electromagnetic units
the force between electric charges e, e' is Kee'/r^2

Units.

and that between magnetic poles m, m' is $\mu mm'/r^2$,
where K, μ are unity in free space. The units of e and m are
modified by Lorentz in such a manner that the forces become
$Kee'/4\pi r^2$ and $\mu mm'/4\pi r^2$. Thus two units of charge in a vacuum
repel with force $1/4\pi r^2$ and the potential due to a charge e in a
medium K is $Ke/4\pi r$. The unit charge of electricity is $1/2\sqrt{\pi}$
times the ordinary electrostatic unit and for Gauss's equation
we have $\int \{\mathbf{dS\, E}\}$ = the surface integral of inward force = $-$ the
total charge inside. Hence if ϕ be the potential,

$$\nabla^2_K \phi + \rho = 0, \quad \left(K \frac{d\phi}{dn}\right)^2_1 + \sigma = 0,$$

and for an electrostatic field,

$$W = \tfrac{1}{2} \int dv \rho \phi + \tfrac{1}{2} \int dS\, \sigma \phi$$

$$= -\tfrac{1}{2} \int dv\, \phi\, \nabla^2_K \phi - \tfrac{1}{2} \int dS \left(K \frac{d\phi}{dn}\right)^2_1 \phi$$

$$= \tfrac{1}{2} \int dv\, K\, \mathbf{E}^2.$$

55. Let us suppose that the unit of length is very small
and that the closeness of examination is such

**The electro-
statics of
stationary
media.**

that the space occupied by an electron may be
considered as finite : we may then suppose that
there is no outer surface of discontinuity bounding
an electron, but that there is gradual transition from the
electron to the empty aether. We shall denote by ρ the
density of the electricity within an electron, and by $\mathbf{e}$, $\mathbf{h}$ the
electric and magnetic forces at any point in this highly magnified

consideration of the conditions. If **u** be the velocity within the electron the Maxwell-Hertzian equations will be

$$\left.\begin{array}{c} \dfrac{d\mathbf{e}}{dt} + \rho\mathbf{u} = V \operatorname{rot} \mathbf{h} \\[2mm] \dfrac{d\mathbf{h}}{dt} = - V \operatorname{rot} \mathbf{e} \end{array}\right\} \quad\dots\dots\dots\dots(40),$$

with $\qquad\qquad \operatorname{div} \mathbf{e} = \rho,$

and on taking the divergence of the former of (40),

$$\frac{d\rho}{dt} + \operatorname{div}(\rho\mathbf{u}) = 0.$$

56. Let us consider a small volume containing a group of electrons for which the total charge, $\int\rho\,dv$, vanishes. We may then call the electric moment of the volume the integral $\int\rho\mathbf{r}\,dv$, where $\mathbf{r}$ is the radius-vector from an origin within the volume to any point where the density is ρ : if $\mathbf{r}'$ is the radius-vector from a second origin whose radius-vector referred to the first origin is $\mathbf{r}_0$, we have $\mathbf{r} = \mathbf{r}_0 + \mathbf{r}'$, and

$$\int\rho\mathbf{r}\,dv = \int\rho\mathbf{r}_0\,dv + \int\rho\mathbf{r}'\,dv = \mathbf{r}_0\int\rho\,dv + \int\rho\mathbf{r}'\,dv = \int\rho\mathbf{r}'\,dv,$$

for $\int\rho\,dv = 0$. Thus the electric moment of the volume so defined is, as it should be, independent of the position of the origin within the volume.

Further, the time-change of the moment will be $\int\rho\dot{\mathbf{r}}\,dv$, or $\int\rho\mathbf{u}\,dv$, the value of which per unit volume is the electric current due to the motion of the charges.

57. If then in the dielectric the electric moment per unit volume, when averaged over an element of volume containing many groups of electrons, is **D**$'$, and if this is changing at a rate $\dot{\mathbf{D}}'$, the current due to this will be $\dot{\mathbf{D}}'$. The equations (40) of the field, which referred to dimensions small compared with the dimensions of an electron, may now be averaged over an element of volume of the size usual in mathematical physics, i.e. containing many groups of electrons : the result is

$$\left.\begin{array}{c} \dfrac{d}{dt}(\mathbf{E} + \mathbf{D}') = V \operatorname{rot} \mathbf{H} \\[2mm] \dfrac{d\mathbf{H}}{dt} = - V \operatorname{rot} \mathbf{E} \end{array}\right\} \quad\dots\dots\dots\dots(41).$$

If the specific inductive capacity be K, and the total polarisation $(E + D')$ be denoted, as before, by D, we know, by comparison with Maxwell's theory, that $D = KE$, and so $D' = (K - 1) E$: this is closely analogous with the corresponding total magnetic polarisation

$$B = H + (\mu - 1) H,$$

where $(\mu - 1) H$ is in these units kH, the moment per unit volume of the induced magnetism, k being the susceptibility.

The equation of continuity for a volume containing a number of groups of electrons is

$$\int dv \frac{d\rho}{dt} = \text{rate of flow of electricity into the volume}$$

$$= \int \{dS . \dot{D}'\} = - \int dv \, \text{div} \, \dot{D}' ;$$

hence, as the volume is arbitrary,

$$\frac{d\rho}{dt} = - \text{div} \frac{dD'}{dt} .$$

Now let us integrate with respect to the time, and remember that the dielectric was uncharged at the time when $D' = 0$ and there were no electric forces,

$$\therefore \ \rho = - \text{div} \, D'.$$

58. This theorem is the analogue of the corresponding expression (13) for the density in magnetostatics and may be stated in the following manner :

If over each molecule or group of electrons $\int \rho \, dv = 0$, then over a region, taken at random and large enough to contain a large number of molecules or groups, the charge per unit volume is $- \text{div} \, D'$, where D' is the mean value per unit volume of $\int \rho r \, dv$.

It is clear that if the boundary were drawn deliberately, with infinite precision, in such a manner as never to cut through any group and so to contain only entire groups, the total charge and so the density would be zero. But when we speak of $- \text{div} \, D'$ as the density we mean the density in any element of volume taken at random.

Regarding this as a purely analytical theorem its application may be generalised in the following manner. If ϕ be any scalar quantity such that $\int \phi \, dv$ vanishes over each group of electrons, then the mean value of ϕ over an element of volume taken at random and containing a large number of groups is $-\operatorname{div} A$, where A is the mean value per unit volume of the integral $\int dv \phi \mathbf{r}$.

59. Let us assume that over a small volume containing

Polarisation of a stationary magnetic medium.

one group of electrons $\int dv \rho = 0$, and $\int dv \rho \mathbf{r} = 0$, where $\mathbf{r} = (x, y, z)$; so that the medium has no electric charge or polarisation. Further, let us assume that

$$\int dv \rho x^2, \qquad \int dv \rho y^2, \qquad \int dv \rho z^2,$$
$$\int dv \rho yz, \qquad \int dv \rho zx, \qquad \int dv \rho xy$$

are all independent of the time. We shall then denote by $\mathbf{m}$ or (p, q, r) the integral $\dfrac{1}{2V}\displaystyle\int dv \rho \,[\mathbf{r}\dot{\mathbf{r}}]$, where the region of integration includes the group of electrons: we shall later see that $\mathbf{m}$ is the magnetic moment of the group. In virtue of the assumptions just made we shall have zero time rates of the quantities $\int dv \rho x^2$, &c., so that

$$\int dv \rho \dot{x}x = 0, \quad \int dv \rho (\dot{x}y + x\dot{y}) = 0; \text{ &c.}$$

Thus
$$\int dv \rho \dot{x}y = \tfrac{1}{2}\int dv \rho (\dot{x}y - x\dot{y}) = -Vr,$$
$$\int dv \rho \dot{x}z = \tfrac{1}{2}\int dv \rho (\dot{x}z - x\dot{z}) = Vq.$$

Further we have, on substituting $\rho\dot{x}$ in the generalised theorem of the previous section, that the mean current parallel to OX, i.e. the value per unit volume of $\int dv \rho \dot{x}$, is equal to

$$-\frac{dP}{dx} - \frac{dQ}{dy} - \frac{dR}{dz},$$

where P, Q, R are the values per unit volume of

$$\int dv \rho \dot{x}x, \quad \int dv \rho \dot{x}y, \quad \int dv \rho \dot{x}z.$$

From any group of electrons the contribution will be

$$-\frac{d}{dy}(-Vr) - \frac{d}{dz}(Vq) \text{ or } V\left(\frac{dr}{dy} - \frac{dq}{dz}\right);$$

and so the component current parallel to OX per unit area will be $V\left(\dfrac{dC}{dy}-\dfrac{dB}{dz}\right)$, where A, B, C are the sum totals per unit volume of p, q, r. Thus if we denote (A, B, C) by G', the electric current, expressed as a vector, is $V\,\mathrm{rot}\,\mathsf{G}'$.

60. Let us consider a stationary medium in which there

Electro-magnetic equations for a stationary medium. is electric density ρ in addition to, and apart from, any effect of the polarisation D', and also a conduction current C. The equations (41) will now become

$$\left.\begin{aligned}\frac{d}{dt}(\mathsf{E}+\mathsf{D}')+V\,\mathrm{rot}\,\mathsf{G}'+\mathsf{C}=V\,\mathrm{rot}\,\mathsf{H}\\[2mm]\frac{d}{dt}(\mathsf{H})=-V\,\mathrm{rot}\,\mathsf{E}\end{aligned}\right\}\ \ldots\ldots(42),$$

with

$$\mathrm{div}\,\mathsf{E}=\text{total densities}=-\,\mathrm{div}\,\mathsf{D}'+\rho,\ \text{i.e. }\mathrm{div}\,(\mathsf{E}+\mathsf{D}')=\rho;$$

and $\mathrm{div}\,\mathsf{H}=0$, for we have no strictly magnetic matter, having merely electrons.

Let us now introduce a new quantity H_1 defined by

$$\mathsf{H}=\mathsf{H}_1+\mathsf{G}';$$

then the equations become

$$\left.\begin{aligned}\frac{d}{dt}(\mathsf{E}+\mathsf{D}')+\mathsf{C}=V\,\mathrm{rot}\,\mathsf{H}_1\\[2mm]\frac{d}{dt}(\mathsf{H}_1+\mathsf{G}')=-V\,\mathrm{rot}\,\mathsf{E}\end{aligned}\right\}\ \ldots\ldots\ldots(43),$$

with $\qquad \mathrm{div}\,(\mathsf{E}+\mathsf{D}')=\rho,\quad \mathrm{div}\,(\mathsf{H}_1+\mathsf{G}')=0\ldots\ldots\ldots(44).$

But these are the ordinary Maxwell-Hertz equations in these units, H_1 being Maxwell's magnetic force and G' the magnetic moment; H or $(\mathsf{H}_1+\mathsf{G}')$ is Maxwell's magnetic induction. We have therefore justified the interpretation of $\dfrac{1}{2V}\displaystyle\int dv\rho\,[r\dot{r}]$,

or $\dfrac{1}{2V}$ (moment of electric momentum) as the magnetic moment of the group of electrons.

61. Returning to the minute scale of examination adopted in § 55, the force acting upon a stationary charge of volume density ρ when the electric force is e will be ρe: if the charge be moving with velocity u the electric force acting upon it will be

Energy in the field and the Poynting flow.

$$e + \frac{1}{V}[uh],$$

which we shall denote by f, and the ponderomotive force per unit volume will be ρ times this, or ρf.

Now over any stationary volume

$$\frac{d}{dt}\int dv\, \tfrac{1}{2}\,(e^2 + h^2) = \int dv \left(e\,\frac{de}{dt} + h\,\frac{dh}{dt}\right)$$

$$\doteq \int dv\, \{e\,(V\operatorname{rot} h - \rho u) - h\,V\operatorname{rot} e\},\ \text{by (40)},$$

$$= V\int dv\,(e\operatorname{rot} h - h\operatorname{rot} e) - \int dv\rho\, eu$$

$$= -V\int dv\operatorname{div}[eh] - \int dv\rho\, uf,$$

by (9), and because

$$uf = u\left(e + \frac{1}{V}[uh]\right) = ue,$$

$$\therefore\ \frac{d}{dt}\int dv\, \tfrac{1}{2}\,(e^2 + h^2) = V\int\{dS\,[eh]\} - \int dv\rho\, uf.$$

Thus the rate at which potential energy increases in any volume, after providing for the rate at which work is done on the moving charges, is equal to the rate of flow into that volume of the vector $V[eh]$ across the bounding surface. This quantity is the Poynting flow of energy and may be denoted by p.

62. Let us now examine the question of whether the interpretation in § 60 of magnetisation as due to the movements of electrons in small orbits will give the same ponderomotive forces as the ordinary theory of magnetic matter.

Forces in a magneto-static field.

Continuing with the minute scale of examination of the previous section we have as the resultant force acting on a group of electrons $\int dv\rho\left(e + \frac{1}{V}[\dot{r}h]\right)$, where r is, as before, the vector to the point from an origin within the group. Now we are here concerned not with the electric and magnetic forces due to the other electrons of the group but with those due to an external field which we may denote by E, H. If E, H be used to denote the forces at the origin those at the point r will be, to the first approximation, $E + r\nabla . E$, $H + r\nabla . H$ and the resultant force on the group will be

$$\int dv\rho \left(E + r\nabla . E + \frac{1}{V}[\dot{r}, H + r\nabla . H] \right).$$

Now $\int dv\rho = 0$, $\int dv\rho r = 0$, $\int dv\rho\dot{r} = 0$,

since the magnetic medium is supposed to have no charge and no electric polarisation. Thus the resultant becomes

$$\frac{1}{V}\int dv\rho\,[\dot{r}, r\nabla . H] \text{ or } \frac{1}{V}\int dv\rho\,[\dot{r}.r\nabla, H].$$

Now we saw in § 59 that

$$\int dv\rho\dot{x}x = 0, \quad \int dv\rho\dot{x}y = -Vr, \quad \int dv\rho\dot{x}z = Vq,$$

so that the operator

$$\int dv\rho\dot{x}.r\nabla = V\left(q\frac{d}{dz} - r\frac{d}{dy}\right),$$

and accordingly

$$\int dv\rho\dot{r}\,\{r\nabla\} = \int dv\rho\,(i\dot{x} + j\dot{y} + k\dot{z})\,\{r\nabla\}$$

$$= V\begin{vmatrix} i & j & k \\ p & q & r \\ \dfrac{d}{dx} & \dfrac{d}{dy} & \dfrac{d}{dz} \end{vmatrix}$$

$$= V[m\nabla] \quad\dotfill(45).$$

Hence

$$\frac{1}{V}\int dv\rho\,[\dot{r}.r\nabla, H] = [[m\nabla]H]$$

$$= \nabla.mH - m.\nabla H$$

$$= \nabla.mH,$$

for div $H = 0$, as the H is due to an external field.

Now the force according to Maxwell's theory is $m\nabla . H$ and

$$m\nabla . H - \nabla.mH = -[m[\nabla H]], \text{ by (2)},$$

$$= 0,$$

for $\qquad$ $[\nabla H] = \text{rot } H = 0,$

as H is due to currents or magnetisation away from the origin.
Hence the resultant force is $m\nabla \cdot H$ as in Maxwell's theory.

63. The couple acting on the group of the previous

Couples in a magneto-static field.
section will be, omitting terms in E which obviously vanish,

$$\int dv\rho \left[\mathbf{r}, \frac{1}{V} \left[\dot{\mathbf{r}}, (H + \mathbf{r}\nabla \cdot H) \right] \right] \qquad \dots\dots\dots(46).$$

Now $\quad \int dv\rho \left[\mathbf{r}, \frac{1}{V} [\dot{\mathbf{r}}H] \right] = \frac{1}{V} \int dv\rho (\dot{\mathbf{r}} \cdot \mathbf{r}H - \mathbf{r}\dot{\mathbf{r}} \cdot H),$

and as in (45),

$$\int dv\rho \dot{\mathbf{r}} \{ \mathbf{r}H \} = V [mH],$$

while $\int dv\rho \dot{\mathbf{r}}\mathbf{r} = 0$, for $\int dv\rho x\dot{x} = 0$, &c.

Hence $\qquad \int dv\rho \left[\mathbf{r}, \frac{1}{V} [\dot{\mathbf{r}}H] \right] = [mH],$

and the contribution to the couple from $\mathbf{r}\nabla \cdot H$ in $(H + \mathbf{r}\nabla \cdot H)$ of (46) will, since $\mathbf{r}$ is very small, be negligible in comparison with that from H which we have evaluated. Thus the couple is $[mH]$ as in Maxwell's theory.

64. The force acting on any particle whose charge is q is

Stresses in the field.
$q\mathbf{f}$, and the stress inside a material medium made up of such particles will be determined by the ordinary laws of mechanics, just as the stresses due to gravity are determined by such forces as mg upon particles of mass m.

Inasmuch as we have no means of measuring stresses in the aether it does not appear that much is gained by obtaining stresses in the aether to explain the force $q\mathbf{f}$ exerted upon particles imbedded in it. There is however some interest in such an interpretation of the phenomena and we may proceed as follows.

65. The resultant force exerted over any volume is

$$\int dv\rho \left(\mathbf{e} + \frac{1}{V} [\mathbf{uh}] \right),$$

3—5

or, as $\rho = \operatorname{div} \mathbf{e}$ and $\dfrac{d\mathbf{e}}{dt} + \rho\mathbf{u} = V \operatorname{rot} \mathbf{h}$,

$$\int dv \left(\mathbf{e} \operatorname{div} \mathbf{e} + \frac{1}{V}\left[\left(V \operatorname{rot} \mathbf{h} - \frac{d\mathbf{e}}{dt} \right), \mathbf{h} \right] \right).$$

On adding to this the contribution

$$- \frac{1}{V} \int dv \left[\mathbf{e}, \left(\frac{d\mathbf{h}}{dt} + V \operatorname{rot} \mathbf{e} \right) \right] + \int dv \, \mathbf{h} \operatorname{div} \mathbf{h},$$

which vanishes in virtue of the second equation of (40) and $\operatorname{div} \mathbf{h} = 0$, we obtain

$$\int dv \, (\mathbf{e} \operatorname{div} \mathbf{e} + \mathbf{h} \operatorname{div} \mathbf{h} - [\mathbf{e}, \operatorname{rot} \mathbf{e}] - [\mathbf{h}, \operatorname{rot} \mathbf{h}])$$
$$- \frac{1}{V} \int dv \left(\left[\frac{d\mathbf{e}}{dt}, \mathbf{h} \right] + \left[\mathbf{e}, \frac{d\mathbf{h}}{dt} \right] \right).$$

Now by (2),

$$[\mathbf{e}, \operatorname{rot} \mathbf{e}] = [\mathbf{e}[\nabla\mathbf{e}]] = \nabla_1 . \mathbf{e}\mathbf{e} - \mathbf{e}\nabla . \mathbf{e},$$

where the suffix of the operator ∇_1 indicates that it acts upon the first $\mathbf{e}$ of $\mathbf{e}\mathbf{e}$ only;

$$\therefore \ [\mathbf{e}, \operatorname{rot} \mathbf{e}] = \tfrac{1}{2}\nabla . \mathbf{e}^2 - \mathbf{e}\nabla . \mathbf{e}.$$

Hence

$$\int dv \, (\mathbf{e} \operatorname{div} \mathbf{e} - [\mathbf{e}, \operatorname{rot} \mathbf{e}]) = \int dv \, (\mathbf{e} . \nabla\mathbf{e} + \mathbf{e}\nabla . \mathbf{e} - \tfrac{1}{2}\nabla . \mathbf{e}^2)$$
$$= \int dv \, (\nabla\mathbf{e} . \mathbf{e} - \tfrac{1}{2}\nabla . \mathbf{e}^2),$$

the ∇ operating in the usual manner on the terms which follow it,

$$= - \int dS \, (\mathbf{N} . \mathbf{e} - \tfrac{1}{2}\mathbf{N} . \mathbf{e}^2),$$

where $\mathbf{N}$ is, as before, a unit vector along the inward normal.

Thus the force is $\mathbf{F}_1 + \mathbf{F}_2$, where

$$\mathbf{F}_1 = - \int dS \, (\mathbf{N} . \mathbf{e} - \tfrac{1}{2}\mathbf{N} . \mathbf{e}^2 + \mathbf{N}\mathbf{h} . \mathbf{h} - \tfrac{1}{2}\mathbf{N} . \mathbf{h}^2),$$

$$\mathbf{F}_2 = - \frac{1}{V} \int dv \, \frac{d}{dt} [\mathbf{e}\mathbf{h}]$$

$$= - \frac{1}{V^2} \int dv \, \frac{d\mathbf{p}}{dt},$$

$\mathbf{p}$ being $V[\mathbf{e}\mathbf{h}]$, the Poynting flow.

66. Accordingly if the field be purely electrostatic, and the components of $\mathbf{e}$ are (X, Y, Z), we shall have $\mathbf{F}_2 = 0$ and the force in the direction of OX may be interpreted as due to a force per unit area upon the surface amounting to

$$- (lX + mY + nZ) X + \tfrac{1}{2} l (X^2 + Y^2 + Z^2),$$

where $\qquad\qquad \mathbf{N} = (l, m, n).$

Thus on a rectangular parallelepiped with its edges parallel to the planes of reference we have forces per unit area parallel to the axis OX:

(a) on the face parallel to YOZ, for which
$$\mathbf{N} = (-1, 0, 0), \ \tfrac{1}{2} (X^2 - Y^2 - Z^2);$$

(b) on the face parallel to ZOX, for which
$$\mathbf{N} = (0, -1, 0), \ XY;$$

(c) on the face parallel to XOY, for which
$$\mathbf{N} = (0, 0, -1), \ ZX.$$

We have accordingly tensions $\tfrac{1}{2} e^2$ along lines of force and pressures $\tfrac{1}{2} e^2$ at right angles to them. This system agrees with Maxwell's stresses in the aether.

CHAPTER VII.

THE ELECTRON THEORY OF LORENTZ APPLIED TO MOVING MEDIA.

67. LET us first of all consider a medium capable of electric but not of magnetic polarisation. Let us take an ordinary element of volume ω containing a number of groups of electrons, and within that element an origin moving with the velocity u of the medium. Then for each group the integral $\int dv \rho = 0$, while $\int dv \rho \mathbf{r}$, the electric moment, is the contribution towards $\omega \mathbf{D}'$, the moment of the element of volume ω.

Case of a non-conducting medium with electric polarisation but with no magnetic susceptibility.

Now since the algebraic sum of the charges in any group is zero we may suppose that when $\mathbf{D}'$ is zero all the electrons in a group are superposed at the origin within the group, and that the polarisation of the medium is effected by moving these electrons from the origin to their final positions. Owing to this movement only, $\dfrac{1}{\omega} \int dv \rho \dot{x}$, the component of $\dot{\mathbf{D}}'$ along OX, is, as in § 56, the rate of flow of electricity per unit area across a plane perpendicular to OX; and hence $\dfrac{1}{\omega} \int dv \rho x$, the component of $\mathbf{D}'$ along OX, may be regarded as the quantity of electricity which has flowed through a plane perpendicular to OX, per unit area.

Thus $\mathbf{D}'$ is a vector of the type considered in § 42, and its rate of change through a circuit moving with the velocity u of the medium is made up of

$$u \ \text{div} \ \mathbf{D}' + \text{rot} \ [\mathbf{D}'u],$$

owing to the velocity of the medium, and $\dfrac{d\mathbf{D}'}{dt}$ owing to time changes independent of that.

Now, since the coordinate of an electron of charge q is $\mathbf{r}$ referred to an origin which moves with the medium, the convection current due to the electron will be $q\dot{\mathbf{r}}$ relative to that origin. If then we take a circuit whose linear dimensions are of the first order of small quantities and which has at each point the velocity $\mathbf{u}$ of the medium at that point, the flow of electricity through the circuit will be that obtained by summing over it the effects of such terms as $q\dot{\mathbf{r}}$. On the other hand the surface integral of $\mathbf{D}'$ over the circuit will be changing at a rate which is due to the same terms $q\dot{\mathbf{r}}$. Thus the flow of electricity through a circuit moving at each point with the velocity $\mathbf{u}$ of the medium will, as when the medium is stationary, be equal to the rate of change of the surface integral of $\mathbf{D}'$ over the circuit: the flow will, accordingly, be

$$\frac{d\mathbf{D}'}{dt} + \mathbf{u} \text{ div } \mathbf{D}' + \text{rot } [\mathbf{D}'\mathbf{u}].$$

Further, as in § 43, the rate of change of the surface integral of $\mathbf{E}$ over the moving circuit per unit area will be the component perpendicular to it of

$$\frac{d\mathbf{E}}{dt} + \mathbf{u} \text{ div } \mathbf{E} + \text{rot } [\mathbf{E}\mathbf{u}].$$

68.   Let us now consider the equations obtained by applying the two fundamental relations to a circuit moving with the velocity $\mathbf{u}$ of the medium. If $\mathbf{E}'$, $\mathbf{H}'$ be the electric and magnetic forces at a point moving with the medium we have

$$\frac{d\mathbf{E}}{dt} + \mathbf{u} \text{ div } \mathbf{E} + \text{rot} [\mathbf{E}\mathbf{u}] + \frac{d\mathbf{D}'}{dt} + \mathbf{u} \text{ div } \mathbf{D}' + \text{rot} [\mathbf{D}'\mathbf{u}]$$
$$= V \text{ rot } \mathbf{H}',$$

and similarly,

$$\frac{d\mathbf{H}}{dt} + \mathbf{u} \text{ div } \mathbf{H} + \text{rot} [\mathbf{H}\mathbf{u}] \qquad = - V \text{ rot } \mathbf{E}'$$

$$\left.\right\} (47).$$

These equations may be written in the form

$$\frac{d\mathbf{D}}{dt'} + \mathbf{D} \text{ div } \mathbf{u} - \mathbf{D}\nabla.\mathbf{u} = V \text{ rot } \mathbf{H}'$$
$$\frac{d\mathbf{H}}{dt'} + \mathbf{H} \text{ div } \mathbf{u} - \mathbf{H}\nabla.\mathbf{u} = - V \text{ rot } \mathbf{E}'$$

$$\left.\right\} \quad \ldots\ldots(48).$$

69. Owing to the magnetisation of moment $\mathbf{G}'$ which resulted from the movement of electrons we had, when the medium was stationary, an electric convection current of amount V rot $\mathbf{G}'$, and the magnetic moment of a group of electrons was the integral through it of $\dfrac{1}{2V}\displaystyle\int dv\rho\,[\mathbf{r\dot{r}}]$. If the medium be moving instead

A moving magnetic medium.

of stationary and the circuit through which the flow is considered be either stationary or moving with the medium, the electric current due to the magnetisation will be still V rot $\mathbf{G}'$: for owing to its physical dimensions a term due to the convection of electrons with velocity $\mathbf{\dot{r}}$ relative to a medium whose velocity is $\mathbf{u}$ may involve either $\mathbf{u}$ or $\mathbf{\dot{r}}$ to the first power, or differentials of these with respect to the coordinates, but cannot involve squares or products of $\mathbf{u}$ and $\mathbf{\dot{r}}$. We may thus equate the current to $\phi\,(\mathbf{\dot{r}}) + \psi\,(\mathbf{u})$, where ϕ, ψ are linear operators. Now putting $u = 0$ we have

$$\phi\,(\mathbf{\dot{r}}) = V\,\text{rot}\,\mathbf{G}';$$

and $\psi\,(\mathbf{u}) = 0$, for it is zero when $\mathbf{\dot{r}} = 0$ and there is no magnetisation. Thus the current in any case is V rot $\mathbf{G}'$.

70. If then we consider the general case in which we superpose the effects of electric and magnetic polarisation, together with a conduction current $\mathbf{C}$, we have

General case of motion.

$$\left.\begin{array}{l} \dfrac{d\mathbf{D}}{dt'} + \mathbf{D}\,\text{div}\,\mathbf{u} - \mathbf{D}\nabla.\mathbf{u} + V\,\text{rot}\,\mathbf{G}' + \mathbf{C} = V\,\text{rot}\,\mathbf{H}' \\[3mm] \dfrac{d\mathbf{H}}{dt'} + \mathbf{H}\,\text{div}\,\mathbf{u} - \mathbf{H}\nabla.\mathbf{u} \qquad\qquad = -V\,\text{rot}\,\mathbf{E}' \end{array}\right\}\;\ldots(49),$$

with

$$\text{div}\,\mathbf{D} = \rho, \quad \text{div}\,\mathbf{H} = 0.$$

Further, on replacing $\mathbf{H}' - \mathbf{G}'$ by $\mathbf{H_1}'$, we obtain

$$\left.\begin{array}{l} \dfrac{d\mathbf{D}}{dt'} + \mathbf{D}\,\text{div}\,\mathbf{u} - \mathbf{D}\nabla.\mathbf{u} + \mathbf{C} = V\,\text{rot}\,\mathbf{H_1}' \\[3mm] \dfrac{d\mathbf{H}}{dt'} + \mathbf{H}\,\text{div}\,\mathbf{u} - \mathbf{H}\nabla.\mathbf{u} \quad = -V\,\text{rot}\,\mathbf{E}' \end{array}\right\}\;\ldots(50).$$

71. If we treat the surface of discontinuity as the limiting
Boundary case of a thin region of continuous transition we
conditions. find, as before, the velocity being continuous, that
the tangential components of E', H_1' are continuous, or

$$[NE']_2^1 = 0, \quad [NH_1']_2^1 = 0.$$

72. Let us determine the result of applying the two
Equations fundamental relations to a stationary circuit
obtained instead of to one moving with the medium.
from a
stationary Let us consider, as in § 42, the cylindrical
circuit. element of volume δv whose ends δS, $\delta S'$ are
formed by the area δS at the times t and $(t + \delta t)$ respectively.
If we regard δS as fixed and $\delta S'$ as moving with the medium,
having started from δS at the time t, the electrons which have
crossed the fixed surface δS during the time δt will be made
up of those which are in the volume δv together with those
which have escaped during the time δt through the moving
surface $\delta S'$ or the tubular surface. The convection currents
or rates of flow at the surfaces δS, $\delta S'$ and the tubular surface
being denoted by F, F', P respectively, the total charge which
has flowed into the volume δv across the ends will, as in § 42,
be $\delta t\, F\delta S - \delta t\, F'\delta S'$; the volume density inside δv being
$- \operatorname{div} D'$ the total charge then will be $- \delta v \operatorname{div} D'$ or
$- \{\delta S\, u\, \delta t\} \operatorname{div} D'$. On comparison with § 42 it will be seen
that the flow in the time δt due to P will be of the order
$(\delta t)^2\, \delta S$, for the quantity which has flowed across the surface
per unit area will be $P\delta t$, and the area itself is of length $u\, \delta t$;
hence the contribution from the tubular surface is negligible by
comparison with the other terms. The resulting equation is
thus

$$\delta t\, F\, \delta S - \delta t\, F'\, \delta S' = - \delta t \,.\, u\, \delta S \operatorname{div} D'.$$

Now $\delta S'$ will differ by δS by quantities of the order $\delta S\, \delta t$
and, omitting quantities infinitely small by comparison with
those retained, we may replace $\delta S'$ by δS in this equation:
thus the flow F through a fixed surface is connected with the
flow F' through a moving surface by the relation

$$F = F' - u \operatorname{div} D' \quad\ldots\ldots\ldots\ldots\ldots\ldots(51).$$

Expressed in non-analytical language the difference between F' and F consists merely in the convection current due to the volume density $-$ div D'.

Now $$F' = \frac{dD'}{dt} + u \text{ div } D' + \text{rot } [D'u]$$

and so, by (51),

$$F = \frac{dD'}{dt} + \text{rot } [D'u] \quad\quad\quad\quad\quad\quad (52).$$

If there be a volume density ρ of electricity in addition to any effects of the polarisation D', we shall have due to that alone

$$F = F' + \rho u \quad\quad\quad\quad\quad\quad (53).$$

A magnetic polarisation due to electrons will have, by § 59, no volume density of electricity, and so will not give rise to any difference between F and F'. Hence, using (52) and (53) instead of (49), we shall have, as the general equations referred to a fixed origin,

$$\left.\begin{aligned}\frac{dE}{dt} + \frac{dD'}{dt} + \text{rot } [D'u] + \rho u + V\text{rot } G' + C &= V\text{rot } H\\[2mm]\frac{dH}{dt} &= - V\text{rot } E\end{aligned}\right\}\dots(54).$$

On replacing H by $(H_1 + G')$ we get

$$\left.\begin{aligned}\frac{dD}{dt} + \text{rot } [D'u] + u \text{ div } D + C &= V \text{ rot } H_1\\[2mm]\frac{d}{dt}(H_1 + G') &= - V \text{ rot } E\end{aligned}\right\}\dots(55).$$

73. If we subtract (54) from (49) we get, using $\rho = \text{div } D$, $0 = \text{div } H$,

$$\left.\begin{aligned}u\nabla . D + D \text{ div } u - D\nabla . u - u \text{ div } D - \text{rot } [D'u]\\= V\text{rot } (H' - H)\\u\nabla . H + H \text{ div } u - H\nabla . u - u \text{ div } H = V\text{rot } (E - E')\end{aligned}\right\}.$$

Hence by (10)

$$\left.\begin{aligned}\text{rot } [Eu] &= V\text{rot } (H' - H)\\\text{rot } [Hu] &= - V\text{rot } (E' - E)\end{aligned}\right\}$$

and we have, as before,

$$E' = E + \frac{1}{V}[uH]$$
$$H' = H + \frac{1}{V}[Eu]$$
...............(56).

Hence also
$$H_1' = H_1 + \frac{1}{V}[Eu]$$ ....(57).

74. We shall first of all consider a slight extension of
Lorentz's transformation theorem. Let us suppose
Effects of motion through the aether. that material media are moving through the
aether with constant velocity u, a function
neither of the coordinates nor the time. The
equations referred to moving axes will be, if there are no
media with magnetic susceptibility, so that $H_1 = H$, and
$H_1' = H'$,

$$\frac{dD}{dt'} + C = V \text{ rot } H'$$
$$\frac{dH}{dt'} = - V \text{ rot } E'$$
...............(58),

where $D = E + (K-1)E'$, div $D = \rho$, div $H = 0$,

$$E' = E + \frac{1}{V}[uH], \quad H' = H - \frac{1}{V}[uE],$$

and $C = \lambda E'$, λ being the conductivity.

At boundaries $[NE']_1^2 = 0$, $[NH']_1^2 = 0$.

These equations must now be transformed by the introduction of new variables, distinguished by double dashes:

$$x'' \equiv x, \quad y'' \equiv y, \quad z'' \equiv z, \quad t'' \equiv t' - (ux + vy + wz)/V^2.$$

Then we shall have

$$\frac{d}{dt'} = \frac{d}{dt''}, \quad \frac{d}{dx} = \frac{d}{dx''} - \frac{u}{V^2}\frac{d}{dt''}, \quad \frac{d}{dy} = \frac{d}{dy''} - \frac{v}{V^2}\frac{d}{dt''}, \quad \&c.,$$

i.e.
$$\nabla = \nabla'' - \frac{u}{V^2}\frac{d}{dt''}$$ ....(59).

Thus
$$V[\nabla H'] = V[\nabla'' H'] - \frac{1}{V}\left[u\frac{dH'}{dt''}\right],$$

i.e. $V \operatorname{rot} \mathsf{H}' = V \operatorname{rot}'' \mathsf{H}' - \dfrac{1}{V} \dfrac{d}{dt''} [\mathbf{u}\mathsf{H}']$

$$= V \operatorname{rot}'' \mathsf{H}' - \dfrac{d}{dt''} (\mathsf{E}' - \mathsf{E}),$$

if squares of $\mathbf{u}/V$ may be neglected.

Hence the first equation of the field, in (58), becomes

$$\dfrac{d\mathsf{D}}{dt''} + \mathsf{C} = V \operatorname{rot}'' \mathsf{H}' - \dfrac{d}{dt''} (\mathsf{E}' - \mathsf{E}),$$

or $\dfrac{d}{dt''} (K\mathsf{E}') + \mathsf{C} = V \operatorname{rot}'' \mathsf{H}'.$

Similarly the second equation of the field becomes

$$\dfrac{d\mathsf{H}'}{dt''} = - V \operatorname{rot}'' \mathsf{E}'.$$

Let us now transform the equations

$$\operatorname{div} \mathsf{D} = \rho, \quad \operatorname{div} \mathsf{H} = 0.$$

We have, by equations (59),

$$\boldsymbol{\nabla}\mathsf{D} = \boldsymbol{\nabla}''\mathsf{D} - \dfrac{1}{V^2} \left\{ \mathbf{u} \dfrac{d\mathsf{D}}{dt'} \right\},$$

i.e. $\operatorname{div} \mathsf{D} = \operatorname{div}'' \mathsf{D} - \dfrac{1}{V^2} \{ \mathbf{u} (V \operatorname{rot} \mathsf{H}' - \mathsf{C}) \}.$

Now $\operatorname{div} [\mathbf{u}\mathsf{H}'] = \mathsf{H}' \operatorname{rot} \mathbf{u} - \mathbf{u} \operatorname{rot} \mathsf{H}',$

and so $-\dfrac{1}{V} \{ \mathbf{u} \operatorname{rot} \mathsf{H}' \} = \dfrac{1}{V} \operatorname{div} [\mathbf{u}\mathsf{H}']$

$$= \dfrac{1}{V} \operatorname{div}'' [\mathbf{u}\mathsf{H}']$$

if squares of $\mathbf{u}/V$ be neglected,

$$= \operatorname{div}'' (\mathsf{E}' - \mathsf{E}).$$

Hence $\rho = \operatorname{div}'' (\mathsf{D} + \mathsf{E}' - \mathsf{E}) + \dfrac{1}{V^2} \{ \mathbf{u}\mathsf{C} \}$

$$= \operatorname{div}'' (K\mathsf{E}') + \dfrac{1}{V^2} \{ \mathbf{u}\mathsf{C} \}$$

and $\operatorname{div}'' (K\mathsf{E}') = \rho - \dfrac{1}{V^2} \{ \mathbf{u}\mathsf{C} \}.$

Similarly $\operatorname{div}'' \mathsf{H}' = 0.$

As boundary conditions we shall have that $[N\mathsf{E}']_1^2$ and $[N\mathsf{H}']_1^2$ shall vanish.

75. Hence the former equations become, neglecting squares of u/V,

$$\left.\begin{array}{l} \dfrac{d}{dt''}\,(K\mathsf{E}') + \mathsf{C} = V\,\mathrm{rot}''\,\mathsf{H}' \\[2mm] \dfrac{d}{dt''}\,(\mathsf{H}') \qquad = -\,V\,\mathrm{rot}''\,\mathsf{E}' \end{array}\right\},$$

with

$$\mathrm{div}''\,(K\mathsf{E}') = \rho - \frac{1}{V^2}\,\{u\mathsf{C}\},$$

$$\mathrm{div}''\,\mathsf{H}' = 0, \quad \mathsf{C} = \lambda\mathsf{E}'.$$

At boundaries $[N\mathsf{E}']_1^2$ and $[N\mathsf{H}']_1^2$ will vanish.

But these are the ordinary equations for the electric and magnetic forces E', H' of the same distribution of material media when t'' is the time, the media now being stationary and the current being $\lambda\mathsf{E}'$, Ohm's law still holding. Also the usual boundary conditions will be satisfied, and the only change is that the new volume density $\mathrm{div}\,(K\mathsf{E}')$ will be

$$\rho - \frac{1}{V^2}\,\{u\mathsf{C}\}.$$

76. Hence it follows that the path of a ray remains on transformation a possible path of a ray and that if squares of u/V be neglected all optical experiments made with sources of light and apparatus fixed with regard to the earth, which moves through the aether with velocity u, would lead to the same results as if the earth were stationary. Thus such experiments, in which there are no conduction currents to cause a change in ρ, cannot, if squares be neglected, lead to any determination of the value of u.

If however the source of light be outside the earth the effects of the motion will become apparent. We shall consider as an example Airy's 'water telescope' experiment in which the effect of aberration was found to be the same when a telescope tube was filled with water as when it was empty.

For the light coming from a star in the direction n or (l, m, n) the electric and magnetic forces in the free aether may be taken as proportional to $e^{is(Vt+lx+my+nz)}$ or $e^{is(Vt+nr)}$, the axes of reference being fixed in the aether: here $s = 2\pi/$(wave length).

If the origin be now taken at a point moving with the velocity $\mathbf{u}$ of the earth, and x', y', z', t' be the velocity and time referred to the new system, we shall have $t = t'$, $x = x' + ut'$, $y = y' + vt'$, $z = z' + wt'$. The forces will then be proportional to $e^{is(V't'+\mathbf{nr}')}$, where $V' = V + \mathbf{un}$.

Now let us apply the transformation theorem and substitute t'', x'', y'', z'' given by

$$t'' = t' - \mathbf{ur}/V^2, \quad x'' = x, \quad y'' = y, \quad z'' = z.$$

The exponential factor then becomes $e^{is(V't''+V'\mathbf{ur}/V^2+\mathbf{nr}')}$, which we may write as $e^{is''(V''t''+\mathbf{n}''\mathbf{r}'')}$, where

$$\frac{s''}{s'} = \frac{V'}{V''} = \frac{l + uV'/V^2}{l''} = \frac{m + vV'/V^2}{m''} = \frac{n + wV'/V^2}{n''}.$$

Hence, omitting squares of small quantities, each portion is equal to $(1 + 2V'\mathbf{un}/V^2)^{\frac{1}{2}}$ or $1 + V'\mathbf{un}/V^2$ or $1 + \mathbf{un}V$ or V'/V. Hence $V'' = V$, as we should expect from the fact that the equations satisfied are those for axes at rest. Further

$$\frac{l''}{l + u/V} = \frac{m''}{m + v/V} = \frac{n''}{n + w/V}.$$

We know that rays of light in the actual will correspond with rays of light in the transformed system. Hence if we consider the rays which come to a focus at a particular point of the observer's eye in the actual and transformed systems, we find that the rays from the star in the actual system will come to the same focus as those which, if the earth were at rest, would have emanated from a star in the direction (l'', m'', n'').

This is what Airy found, for the direction is that of the resultant of (Vl, Vm, Vn) and of (u, v, w). Further the period is quickened in the ratio of s'' to s or of ($V + \mathbf{un}$) to V, which is in accordance with Döppler's law.